PRINCIPLES OF INSECT BEHAVIOUR

2nd Revised and Enlarged Edition

Glass-wing butterfly *Greta* sp. is overlooked and invisible against flowers and other vegetation due to its transparent wings with very few scales.

PRINCIPLES OF INSECT BEHAVIOUR

2nd Revised and Enlarged Edition

V.B. Awasthi

M.Sc. Ph.D.

Former Associate Professor
Department of Zoology
J.N.V. University, Jodhpur (India)

SCIENTIFIC PUBLISHERS (INDIA)

Published by:

Scientific Publishers (India)
5 A, New Pali Road, P.O. Box 91
Jodhpur 342 001 (India)

Branch Office
Scientific Publishers (India)
4806/24, Ansari Road, Daryaganj
New Delhi - 110 002 (India)

E-mail: info@scientificpub.com
Website: www.scientificpub.com

First Edition, 2001
Second Edition, 2013

ISBN: 978-81-7233-825-1

Typeset by Rajesh Ojha
Printed in India

Preface (1st Edition)

Although there are several books on animal behaviour which mainly deals with vertebrate fauna but there are only a few books on insect behaviour. The information on insect behaviour is very vast, but is scattered in research papers, review articles and monographs. Hence, information from all the available sources has been collected and compiled. The principal aim of writing this book has been to provide in short, the relevant information on the basic principles of insect behaviour together with modern trends on this fascinating and advancing subject.

This book is useful to the students of Entomology and to the insect Ethologists engaged in research work because it provides basic as well as advanced information with an exhaustive but selected bibliography which deals with most of the major aspects of the subject. This book is equally useful to the students of Zoology who study insect behaviour along with animal behaviour as a part of their syllabus in most of the Indian Universities.

This book is also useful to the general readers who have some basic knowledge of Biology and have curiosity to know why the insects are vividly coloured, have peculiar shapes, produce sound and light, and how they communicate.

It may be pointed out here that in each act of insect behaviour numerous well known external, internal as well as some little known factors are involved at times, because nature visualizes each pattern of behaviour in totality and not in individuality, hence it was very difficult to single out which one factor is more important for a particular behaviour and one is less. Some factor(s) may be significant or more pronounced in a particular pattern of behaviour, while they may also be of equal importance for controlling other pattern(s) of behaviour. Hence, the readers may find some sort of overlapping here and there in the text. It may be added that for a proper understanding of a particular behaviour pattern which is influenced by more than one factor such sort of overlapping was unavoidable.

It is hoped that this book will be useful not only to the students of Entomology-Zoology and research scholars of insect behaviour but also to naturalists who wish to know about some of the mysteries of these small creatures and their life style. For the sake of convenience the book has been divided into 20 chapters viz —

(1) Kinds of insect behaviour (2) Camouflage, warning colouration and defensive behaviour (3) Social behaviour and parental care (4) The language of bees and their behaviour (5) Interactions between insects and plants as related to pollination (6) Food and feeding behaviour (7) Intraspecific and interspecific interactions (8) Adaptations in insects (9) Thermoregulation in desert insects (10) Photoperiodism and insect behaviour (11) Photoperiodism and ecological adaptations (12) Hormonal control of insect behaviour (13) Pheromonal control of insect behaviour (14) Genetic control of behaviour and evolution of behaviour (15) Communication in insects (16) Insect pest management by behavioural means (17) Migratory and dispersal behaviour of some insects (18) Influence of host plants on sex pheromone behaviour of phytophagous insects (19) The role of sense organs in mating behaviour of cockroaches and (20) Reproductive behaviour.

I would like to mention that a large number of insect Ethologists have helped me directly or indirectly in this endeavour but it is not possible to acknowledge them individually. I express my most humble and profound regards and sincere gratitude to all of them.

Thanks are also due to the authorities of the Scientific Publishers, Jodhpur, Shri Prahlad Rai Sharma and Shri Pawan Kumar Sharma for their encouragement, and keen interest in my work and bringing out the book in an excellent form.

I am also grateful to my friend Shri A.P. Jain, Botanical Survey of India, Jodhpur, for carefully going through the proof and offering valuable suggestions.

Finally, I also wish to thank my family members : Wife Sneh Lata, Son Santosh and daughter Shubha for their whole hearted co-operation without which this work was not possible.

V.B. Awasthi

30th May, 2000
Jodhpur

Preface to the Second Edition

With great pleasure I am presenting the Second edition of my book "Principles of Insect Behaviour". In this edition, a lot of new information has been added in the text at appropriate places alongwith a good number of illustrations. Two new chapters viz. 1. Ecological, behavioural and other adaptations of desert insects and 2. Thermoregulation by behavioural means have been included. Further, three appendixes have also been given dealing with 1. Amazing Insects, 2. Bee facts and 3. A copy of letter of Prof. V.B. Wigglesworth - The Father of Insect Physiology, in response to one of my critical letter which shows his magnanimity. I take this opportunity to express my gratitude to Prof. Wigglesworth for his useful suggestions received from time to time.

I hope the readers will find the book interesting as well as useful and welcome this edition with a great zeal.

V.B. Awasthi

28th November, 2012
Jodhpur

Contents

1

KINDS OF INSECT BEHAVIOUR

Broadly speaking the manner in which an organism responds to various exogenous and endogenous stimuli is termed as behaviour. Insects mainly exhibit two kinds of behaviour viz. (1) Innate behaviour which include reflexes, kineses and taxes and (2) Learned behaviour i.e. experienced behaviour.

The innate behaviour largely consists of more or less fixed response or series of responses to a given stimulus or the pattern of stimuli. It may be added here that the innate behaviour may be modified by experience or learning. The innate behaviour is hereditary whereas the learned behaviour is not inherited but it is acquired through interaction with the environment during the life of the individual. In several instances, there is no line of demarcation whether an observed behaviour is innate or learned. In the following pages a brief account of both the types of behaviour innate and learned is given.

INNATE BEHAVIOUR

The simplest innate behaviours found in insects are reflexes. This may involve only a part of the body as in the case of extension of proboscis or the whole body as in the case of direct reflex which occurs when an insect is placed on its dorsal side. The reflexes can be classified into two groups: 1. phasic and 2. tonic. The *phasic* reflexes are much rapid and short lived and are involved in rapid movements e.g. the extension of proboscis in flies. On the other hand the *tonic* reflexes are slow and long lived and are involved with the maintenance of posture, body turgor, muscle tone and equilibrium. The reflexes also vary in complexity, the simplest being mediated by a single affrent impulse from a receptor to an interneuron and then along an efferent neuron to an effector i.e. the reflex arc. However, most of the reflexes are very complex involving many neural connections. Reflex autonomy has been shown by ganglia of the ventral nerve cord e.g. in the silkmoth, *Bombyx mori* the oviposition reflex which results from the ovipositor making contact with a surface, resides totally on the last abdominal ganglion. More complex innate behaviour includes various types of orientation patterns.

Orientation may be defined as "the capacity and activity of controlling locomotion and attitude in space and time with the help of exogenous (external) and endogenous (internal) stimuli (Jander, 1963). Orientation in insects has been variously classified by different workers. For the sake of convenience, here it has been broadly classified into three types (1) kineses, (2) taxes and (3) transverse orientation which are based on the mechanisms of orientations. Kineses and taxes can be defined with respect of the type of stimulus.

Kineses: The kinesis is an unorientated or undirected action of an organism which varies according to the intensity of stimulus e.g. in *Glossina* species, there is an increase in activity in dry atmosphere relative to activity in a moist one. This behaviour causes the insect to find and remain in a humid atmosphere. The body louse, *Pediculus* makes fewer and fewer directional changes as the temperature, humidity and odour increase as the louse moves closer to a potential host.

In addition to the providing specific environmental input, impulses from sensory neurons also increase overall excitability of the central nervous system turning on the insect as it were e.g. increasing temperature above the preferred range for an insect, generates increased locomotor activity resulting ultimately in the insect's moving into a cooler area. This kinesis may appear to be directed movement, the insect seems to avoid the heat by remaining in motion and turning until in a cooler place (Wigglesworth, 1972).

Orthokinesis: It is defined as a response to a stimulus in which the rate of response such as speed of movement is positively proportional to the intensity of the stimulus.

Klinokinesis: Movement resembling a random walk in which changes in direction are made when unfavourable stimuli are encountered with frequency of turning dependent upon the stimulus intensity.

Akinesis: It is a state of immobility due to lack of any stimulus. An overall lack of movement is controlled by sensory input normally by tarsal contact with a surface. Tarsal contact combined with other stimuli e.g. for dragon fly, lowered temperatures and reduced intensity of light in the evening inhibits central nervous system function in most insects (Wigglesworth, 1972).

Taxes: An oriented or directed movement of an organism towards or away from the stimulus is known as taxis. The taxes are different from the kineses in the sense that they are directed responses relative to a stimulus source. Taxes may be responses to physical or biological factors or a combination of both. The insect may move towards (positive) or away (negative) from the source of stimulus. Thus, moth's attraction to light is positive phototaxis while maggots are negatively phototactic.

The taxes may be classified according to the environmental stimuli involved e.g. phototaxis (light), geotaxis (gravity), anemotaxis (air current), rheotaxis (water current), chemotaxis (chemicals), hygrotaxis (humidity), scototaxis (darkness), thigmotaxis (contact) for details see: Wigglesworth (1965). The other taxes are klinotaxis, menotaxis, telotaxis, and tropotaxis.

Klinotaxis: In klinotaxis behaviour an insect moves relative to a gradient of stimulus intensity such as light source or a sound emission. Thus, a movement in a definite direction relative to a stimulus either directly towards or away from the source is known as klinotaxis.

Menotaxis: Movement that maintains a constant orientation relative to a light source e.g. the orientation with respect to a constant angle of light source is termed menotaxis.

Telotaxis: The visual fixation of an object such as prey is termed telotaxis i.e. orientation and movement directly towards a visual stimulus.

Tropotaxis: Movement whose direction is dependent on the composition of stimulation on bilateral sense organs. Attainment of orientation is directed and no deviations are required. The tropotaxis is more efficient than klinotaxis and shows the selective advantage of having paired sensory structures viz. eyes, cerci, antennae. If an antenna is removed or an eye is painted over, the insect will not orient properly. If antennae are crossed and then glued in their new position, an insect will turn towards the wrong side when orienting to an odour source.

It must be remembered, when attempting to analyze the behaviour of an insect that the mechanism involved may change from moment to moment and even at the same instant more than one mechanism may be involved.

Transverse Orientation

The transverse orientations are those in which the body is oriented at a fixed angle relative to the direction of the stimulus. The transverse orientations are of two types: (1) the light-compass reaction and (2) the dorsal light reaction (Fraenkel and Gunn, 1961). The light compass reaction consists of an orientation such that locomotion occurs at a fixed angle relative to light rays. This orientation was demonstrated in ants, caterpillars, bees, certain beetles and bugs. In the dorsal or ventral light reaction, both the long and transverse axes of the body are kept perpendicular to a directed source of light (Fraenkel and Gunn, 1961). This type of reaction is different from phototaxis because in this case, orientation is transverse to the light rays and not parallel to them e.g. both dorsal and ventral light responses are found in certain aquatic bugs. When the air bubbles trapped between the head and the antennae are removed, the back swimmer, *Notonecta,* depends entirely upon a ventral light reaction for the maintenance of its normal swimming position, while the converse is true for the water boatman, *Corixa* and the creeping water bug, *Naucoris* (Rabe, 1953).

The most complex innate behaviour patterns are the fixed action patterns. These are characterized by being unlearned, species specific and adaptive. They differ from reflexes and orientations in that they require an internal readiness before they can occur and the evoking stimulus i.e. releaser is commonly not required to act during the entire course of the pattern. The releasing stimulus may be internal in the form of hormone secretion or an excitatory signal from a higher neural centre. The stimulus may be in the form of light, temperature, mechanical contact, odour etc. A variety of factors may affect the development of a state of internal readiness or motivation necessary for a given behaviour pattern to occur. These may be either internal or external in origin (Romoser, 1981).

The internal motivational factors are hormonal or neural in nature e.g. the relative concentration of juvenile hormone and the moulting hormone determine the spinning behaviour of the caterpillar of several moths and other insects. Internal receptors are commonly involved when a behaviour pattern appears spontaneously without any discernible external stimulus. An excellent example of this has been found in the blow fly, *Phormia,* in which receptors associated with the crop carry stimuli that inhibit food ingestion as long as the crop is full. No inhibitory stimuli arise from these receptors when the foregut is empty and feeding can occur. In other cases of spontaneous behaviour, regular cycles of behavioural activity of internal origin determine when a particular pattern may be released e.g. crickets are active only during a certain period each day and this daily rhythm continues for several days when the insects are kept in constant darkness in the laboratory.

The external factors viz. light, temperature, substances produced by or activities of other insects of the same species and others have definite effects on motivation, in addition to serving as releasers (stimulus). Light and temperature commonly determine whether an insect is active or inactive. Every insect species has a temperature below which or above which it becomes inactive and within the range of these temperatures; specific temperature may be essential for particular activities e.g. the readiness of certain butterflies, *Pieris* etc. to copulate is determined by temperature. Likewise, the entire act of mating behaviour in the milkweed bug *Lygaeus militaris* depends on temperature. Mating takes place only at a temperature of 38-46°C or above as observed in this bug in the Indian desert (Awasthi and Tewari, 1975).

It is well known that a substance secreted by one insect influences the behaviour of another insect of the same species. The term *Pheromone* is commonly used to denote such a substance. Pheromone may serve as releasers or as motivational factors or both. One of the most interesting example of a pheromone affecting behaviour is found in the honey bee. The queens produce a substance known as queen substance (secretion of the mandibular glands) which is passed from worker to worker orally and which inhibits ovary development as well as the behaviour involved with queen cell construction. Queen cells, in which new queens develop, are constructed only when a hive has lost its queen and hence the effects of the substance of her mandibular glands. Jacobson (1965, 1972, 1974) has described in considerable detail the various aspects of insect pheromones.

LEARNED BEHAVIOUR

According to Thorpe (1963) learning is that process which manifests itself by adaptive changes in individual behaviour as a result of experience. Thus learning involves the accumulation and storage (memory) of environmental information and the subsequent effects of this stored information on the behaviour of an animal, i.e. insect. Alloway (1972) defined learning as a modifiable relatively permanent change in behaviour as a result of practice. The simplest form of learning i.e. habituation is common among insects. Markl (1974) has classified the learning behaviour into (i) habituation, and (ii) latent learning.

Habituation: Different authors have defined this term in their own ways e.g. according to Dewsbury (1978) habituation may be defined as the decrease in probability or amplitude of a response occurring when a stimulus which elicits that response is presented repeatedly.

According to Gullan and Cranston (1994) habituation means reduction in the response to a stimulus with repeated exposure through modification of the central nervous system.

Habituation (getting used to) has occurred when a stimulus that initially evoked an escape or avoidance response no longer evokes such a response. This type of learning is common in insects and has been demonstrated relative to substrate vibrations noise, visual stimulation, chemical repellents etc. The ability of insects to learn in this fashion avoids unnecessary, energy consuming behavioural responses (Romoser, 1981).

Latent learning: It occurs without apparent reward or punishment. Social insects learn characteristics of the immediate vicinity of their nest so that they are able to find their way back from foraging flights. Olfactory conditioning may be a form of habituation or latent learning. An example of this phenomenon is a situation in which exposure to a repellent odour during a certain period of development results in the loss of repellency of this odour to adults e.g. the adults of *Drosophila melanogaster* which are normally repelled by the odour of peppermint oil. However, if the larvae have been reared in the presence of peppermint they are strongly attracted to it as adults.

Instincts: The inborn patterns of movement or automatisms, brought into operation by some appropriate conditions are termed as instincts. Instincts are normally manifested in response to a specific releaser stimulus but in the absence of such releaser, the threshold of stimulation falls and the whole elaborate response may be set off by some quite abnormal circumstances. These inborn patterns of behaviour are genetically determined or inherited, they are built into the central nervous system in the same way as morphological patterns are built into the growing body (see: Wigglesworth, 1965). Darwin (1859) has described slave making instincts and cell making instincts in detail.

Periodicity of Behaviour

Many of the cyclic activities of insects like locomotion, feeding, mating, oviposition, eclosion and others occur at regular intervals. The timing of these recurring activities is adaptive i.e. insects under natural conditions behave in ways that maximize the chances of their survival. Thus phytophagous insects hatch from eggs in synchrony with the availability of host plants, insects become dormant and cold hardy or migrate at a time that favours survival of an upcoming cold or dry season. The term periodicity is applied to such activity cycles of behaviour. The patterns of cyclic activities may be every 24 hours, monthly, or annually in coordination with lunar cycle. Periodicity in insect behaviour involves internal as well as external components. Internal or endogenous components are related to the heredity while external or exogenous components arise from the external environment.

Circadian Rhythms: These are processes that are controlled by biological clock. Most rhythms that have been identified have a period of approximately 24 hrs and are referred to as circadian (circa- about or approximately, diem- a day or daily) rhythms. The term circadian rhythm was coined by Franz Halberg (Halberg *et al.*, 1959). The diurnal species as most bees, wasps, dragonflies and butterflies are those whose activities are limited mainly to day light hours while nocturnal insects i.e. most moths and the cockroaches, crickets are active during darkness.

The crepuscular insects are most active around dusk and dawn e.g. *Drosophila* emerge from the puparium at dawn and likewise several orthopternas sing at a particular time of day or night. Circadian activity rhythms though internally controlled are subject to external modification paticularly by temperature (Truman, 1973).

Numerous non-circadian rhythms are also known e.g. the lunar emergence of some chironomid flies (Diptera), mayflies (Ephemeroptera) and the annual emergence of some dermestid beetles (Corbet, 1966). A well known example of noncircadian thythmicity is found in the species of periodical cicadas viz. *Magicicada* spp. (Homoptera), the various broods of which emerge *en-masse* at 13 or 17 year intervals.

Although rhythms are not regulated by external (exogenous) factors, but some external factors are known to influence rhythms. Among the external factors that may act as time cues are length of the light or dark phase of a photoperiod (i.e. a cycle consisting of a period of illumination followed by a period of relative darkness), light intensity, temperature, time at which food is available etc. (Beck, 1968).

Many exogenous environmental factors display 24 hrs periodicity which may result in corresponding periodicity in insects without the involvement of biological clocks. Among these are - intensity of light, humidity, temperature, wind velocity and numerous biotic factors.

An example of how one of these factors may cause periodic behaviour is in the swarming of many species of mosquitoes, which is determined by the intensity of light. These insects continually possess the drive to swarm but do so only under appropriate conditions of light intensity and can be induced to swarm for hours if the light intensity is held artificially at the appropriate level. Under natural conditions the appropriate light intensity may be periodic e.g. that which occurs during crepuscular periods. In this situation, the mosquitoes would swarm periodically.

CONTROL OF INSECT BEHAVIOUR

Romoser (1981) remarked that an insect is continually attcked with several stimuli in space and time. Many stimuli provide information significant to the life of the insect but many do not. In order to survive, an insect must respond to the right stimuli and do the right thing at the right time escape, locate a food source, feed, locate a mate, copulate, oviposit, migrate, become dormant and so on, i.e. behaviour of an insect must be controlled. The principal factors that control the behaviour of insects are nervous and endocrime systems.

Nervous Control of Insect Behaviour

It is known from behavioural and neurophysiological studies that patterns of behaviour may occur spontaneously without any external sensory triggering. Thus, the nervous system must be capable of spontaneously generating organized patterns of behaviour from within e.g. a decapitated mantid will continuously turn and perform copulatory movements. The central nervous system in a locust with all sensory input from the head and wings removed still produces motor output nearly identical to the motor output recorded during normal flight (Wilson, 1972). These activities are considered to arise from the spontaneous, preprogrammed activity of neural pattern generators in the central nervous system.

The brain and suboesophageal ganglion both have been found to exert excitatory and inhibitory influences on internally (endogenously) controlled motor patterns. These patterns seem to reside largely in the thoracic ganglia in the flight and in this way the decerebrated mantid sends out impulses that inhibit the internal activity of the ganglia of the ventral nerve cord relative to copulation. In the case of locust, there are wind sensitive sensilla on the head which help sustain flight and also tarsal receptors which stimulate flight when contact with the substrate is lost. In addition, there are stretch receptors which aid in the regulation of flight on wing hinges and other parts of the wings and mesothroax as well as metathorax. Thus, although locust flight pattern is centrally generated, the basic pattern is modified by peripheral sensory input. Although a little is known about the neurophysiological basis of learning but removal of parts of the brain has shown that some long term memory must reside there.

Endocrine Control of Insect Behaviour

Hormones play significant roles in insect behaviour (see: Chapter 12). Through acting directly on the nervous system they can cause major overall changes in paterns of behaviour. Endocrine mechanisms typically exert their influence over the longer term, coordinating behaviour with less immediate environmental demands e.g. changing seasons, changes in the availability of food etc. The determination of hormonal influences on behaviour have usually been based on one or more of the following criteria:

1. Correlation of hormone secretion and a particular behaviour pattern. 2. Application of hormone or implantation of active endocrine glands inducing a behavioural pattern sooner than it would normaly occur. 3. Correlation between removal of any active endocrine gland and disappearance of a specific behaviour. 4. Restoration of the specific behaviour by hormone application or implantation of active endocrine glands. It is the last criterion that provides the strongest evidence for hormonal control of a given behaviour. According to Riddiford and Truman (1974) two categories of hormonal effects on behaviour i.e. as modifiers or as releasers of behaviour can be recognized.

Hormones as Modifiers of Behaviour

When a hormone acts as a modifier, it changes the responsiveness of an organism. The example of a hormone acting as modifier is the influence of moulting hormone on the biting behaviour of female *Anopheles freeborni* (Beach, 1979). This mosquito does not take blood meals once ovarian development has been initiated. The moulting hormone, which in this case, is secreted by the ovaries inhibits biting. This was concluded on the basis that biting inhibition does not occur after removal of the ovaries and is restored by either reimplantation of ovaries or injecting moulting hormone (Romoser, 1981).

Hormones as Releasers of Behaviour

When a hormone acts a releaser of behaviour, a specific behaviour pattern occurs within a few minutes of secretion e.g. phalleic nerve stimulating hormone in the *Periplaneta americana* (Milburn *et al.,* 1960; Milburn and Roeder, 1962). This hormone can be extracted from the corpora cardiaca and when injected into male cockroaches causes the abdominal movements characteristics of copulation. The eclosion hormone secreted by the brain (i.e. neurosecretion) and corpora cardiaca in the moth *Antheraea pernyi* acts both as a releaser in triggering pre-eclosion behaviour as well as modified by turning on adult behaviour (Truman, 1973).

2

CAMOUFLAGE, WARNING COLOURATION AND DEFENSIVE BEHAVIOUR

In addition to its hardness, the form and colouration of the insect cuticle offers protection to many insects. Protective resemblance, wherein an insect's colouration and/or appearance apparently deceives predators, abounds in the class insecta. Insects have evolved similarities to thorns, seed heads, sticks, leaves, twigs, stones, bark, bird-droppings, flowers and a host of other inedible items. Here only a brief account of the camouflage, warning colouration and defensive behaviour is given.

CAMOUFLAGE AND WARNING COLOURATION

Considerable variations are displayed by insects in their form, colouration as well as the patterns of colouration. The close resemblance of wings and other parts of the body to leaves (leaf butterflies), *Kallima* spp. (Fig. 1A,B), twigs, sticks (Figs. 2-4) thorns and other parts of the plant like seed heads (Fig. 5), insects with eye spots (Owl butterfly) (Fig. 6) and 10 moths (Fig. 7 A,B) or bright colours on wings which are suddenly displayed in response to a threat of vertebrate predators, learn to avoid through experience with poisonous stings, irritating sprays, disagreeable tastes, warning colouration (Figs. 8,9, 10AB, 11A-C) etc. and insects that blend in so well with their surroundings that they become nearly invisible to predators are a few examples (see: Figs. 12, 13, 14 & 15) of camouflage and warning colouration.

Interestingly enough, most of the silkmoths have prominent round, eye-like markings which make a rather frightening spectacle to a prospective predator. The owl butterfly *Caligo* sp. (Fig. 6) has similar markings which are very effective when it suddenly flips its wings open and closed in the face of a predator. The *Automeris* have the eyes on the hindwings which are concealed until they are disturbed when they are flashed open, displaying brightly coloured target like markings. Such flash colouring is also adapted by the peacock butterfly, *Inachis io* whose brightly coloured wings have camouflage pattern. When disturbed the peacock butterfly opens its wings suddenly and makes a loud rustling sound.

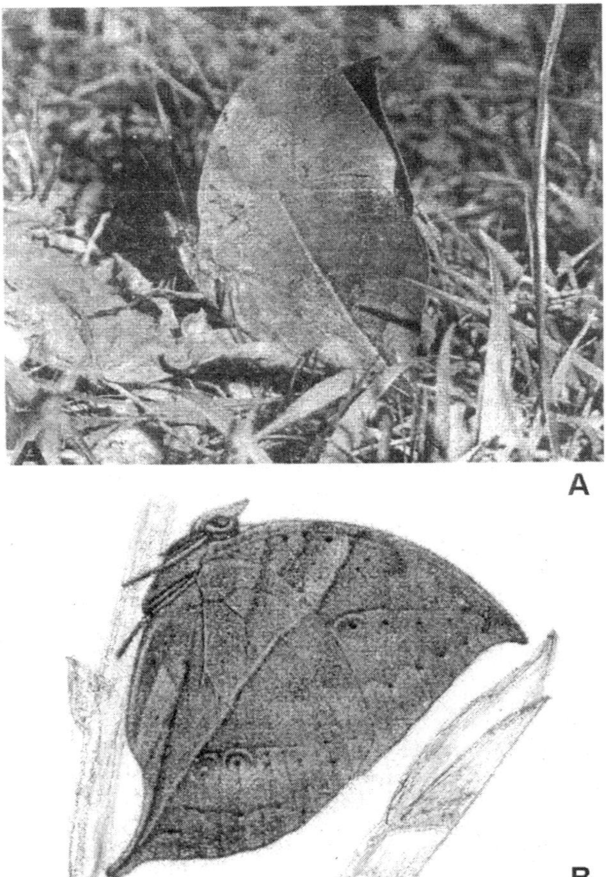

Fig. 1. Leaf butterflies A. *Kallima inachus,* B. *Kallima limborgi*

Fig. 2. Caterpillar of peppered moth appearing like a branch on which it is living

Fig. 3. A Bordered beauty moth caterpillar appearing like a growing twig.

Fig. 4. A walking stick

An insect must behave in an appropriate way if it is to gain any protective benefit from its form and colouration. Thus an insect that closely resembles a leaf will tend to orient itself on a stem in an appropriate manner, one with eye spots or other brightly coloured markings will display these characteristics when threatened by a predator and one that resembles other species may also adopt similar behaviour. Insects that blend with their surroundings tend to remain motionless when threatened. The examples of protective form, colouration and pattern include crypsis, aposematic colouration, mimicry and spots (Figs. 11-18), such protective devices would be of no use without appropriate behaviour.

Fig. 5. Larvae of *Melicta athalia* appear like seed heads of plantain on which they feed.

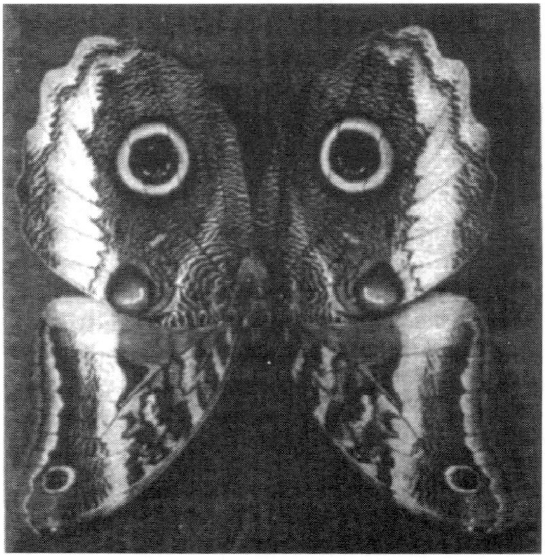

Fig. 6. Owl butterfly, *Caligo* sp. its wing pattern appears like the eyes of owl, due to which it is protected from predators.

The term `crypsis' referes to various combinations of form, colour and pattern that facilitate hiding from predators. Resemblance to various inedible repulsive objects like thorns or bird droppings (of the larva of *Papilio demoleus*), colouration and shading that cause an insect to blend into the background and transparent wings that allow the background to show through all, come under crypsis category.

The term "aposematic" or warning colouration is also some times called `advertising' in the sense that an insect tastes very bad, stings or does something disagreeable to a predator and communicates this ability visually by colours and patterns that contrast blatantly with the background (Figs. 11-18). An aposematically coloured insect may also be advertising falsely in that although it looks like a harmful insect but it is really harmless and palatable (Figs. 11 A-C, 21 A-C). Aposematically coloured insects are usually boldly patterned in shades of orange, red, or yellow contrasting with black and stand out against the greens and browns that are characteristic of environment. Unpalatability is commonly due to the presence of chemicals in the haemolymph that have been derived from host plants e.g. the Monarch butterfly, *Danaus plexippus* and certain true bugs and beetles which feed on various species of milkweed plants (*Calotropis* spp., Asclepiadaceae), accumulate chemicals specially cardiac glycosides from these plants. These chemicals not only render the insect unpalatable to vertebrate predators particularly birds but also act as emetics, i.e. they make a predator to vomit.

The term "mimicry" refers to the close resemblance in form, colour, pattern etc. between one insect (the mimic) and another insect or a plant part (the model) insects may mimic seed heads, thorns, twigs, leaves etc. (see Figs. 1-18) and thus the third animal is deceived into confusing the two. In other words the mimicry has been defined as the resemblance of a mimic to a model by which the mimic derives protection from predation provided to the model (e.g. by unpalatability). Many insects e.g. moths, butterflies and caterpillars, bear spots at various places on their body (Figs. 6,7AB). These spots often closely resemble eyes of vertebrates and in some cases eyes of other insects. Many moths which have eye spots on the hind wings are otherwise cryptically coloured when resting on their usual substrate (Fig. 14 A-C). If the first line of defence, crypsis, fails and a predator strikes, the forewings are thrown forward suddenly displaying the eye spots. Such behaviour has been shown experimentally to frighten and repel insectivorous birds (Blest,

1957). The eye spots (Figs. 6,7 AB) resemble eyes of birds e.g. owls that prey on insectivorous birds.

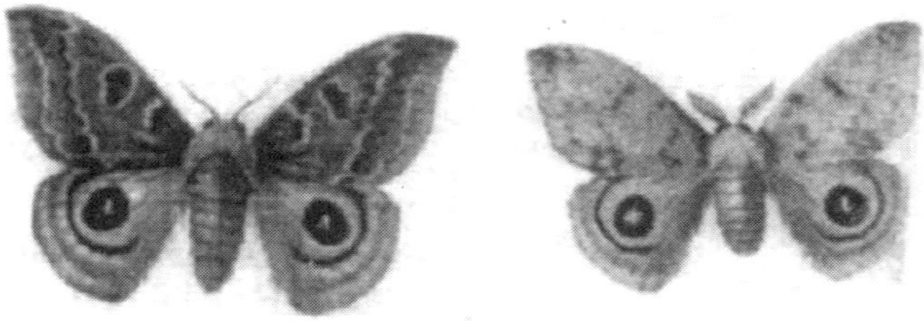

A B

Fig. 7. A. Male and B. Female 10 moths showing eye spots on the hind wings.

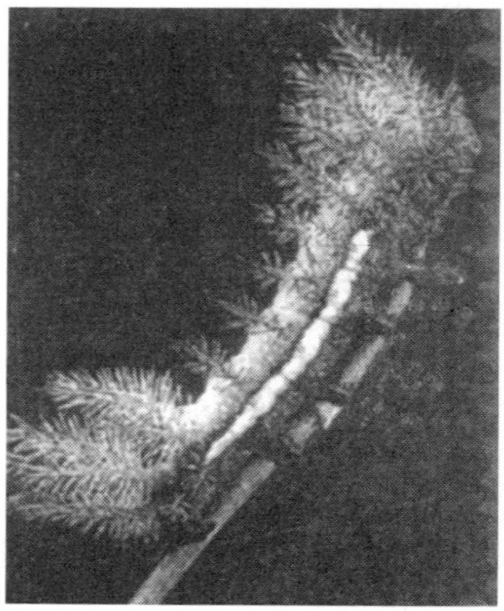

Fig. 8. Caterpillar of a moth bearing bunches of stinging spines that burn and irritate the skin of any one handling them.

Fig. 9. Larva of a moth having warning colouration for predatory birds together with poisonous spines on the body for protection against insect parasites

Fig. 10. **A.** Pussa moth caterpillar adapted for defence. **B.** Saddleback caterpillar with poisonous spines

Fig. 11. **A.** *Vespula vulgaris* (warning colouration) **B.** Wasp beetle, appears wasp-like but harmless, **C.** *Zygaena occitanica* (warning colouration)

It may be pointed out here that the mimic is usually less abundant and exists in the habitat as model, whenever, the mimic becomes more abundant or the model less numerous then selection pressure is increased on both groups. There are two common types of mimicry. The one Batesian mimicry named after Henry W. Bates (1862). In this case the mimic is restricted to those species that are palatable but gain benefit because predators are apparantly deceived e.g. Monarch (model) and the Viceroy (mimic) butterflies (Fig. 16 A,B). Both species have similar colour patterns and birds having learned to avoid Monarch also avoid Viceroy. As remarked by Brower (1969) an untrained bird may feed on numerous viceroy and palatable monarchs (those that have developed on milkweed without toxins) until encountering its first unpalatable model after which further predation upon this colour pattern ceases.

Fig. 12. **A.** Orange tip butterfly, an example of concealing colouration, **B. & C.** from lateral and dorsal views.

The second type of mimicry is Mullerian mimicry named after Fritz Muller (1879). In this case both the model and mimic are unpalatable and the ingestion of either by the vertebrate predators results in the avoidance of both the species. Whereas Batesian mimicry is usually disadvantageous to the model, the Mullerian mimicry results in benefits to both the mimic and the model since both are unpalatable (*vide supra*). Brower (1960), Brower *et al.* (1963) studied the relative palatability of helioconian butterflies to silverbeak tangers (finches) in Trinidad and noted that similar shape and size gave selective advantage to the Mullerian complex of butterflies. In addition to butterflies, the other examples of Mullerian mimicry can be seen among the lycid beetles, wasps, bees and ants. The other two viz. Wasmannian mimicry, and Peckhammian or aggressive mimicry are less common, hence, not dealt in detail.

The basic principle of camouflage is to be coloured like the background. Another step towards perfection in natural camouflage is the development of the desruptive patterns to break up the outline of the animal and hide its shape. This is more important in larger animals but it is often seen in insects which sit on the tree trunks and branches e.g. moths and bugs (Fig. 14 A-C). A bold dark and light pattern combined with the resemblance to background will give more effective concealment than an even tone all over (Figs. 14 A-C, 17, 18). Some short horned grasshoppers (Acrididae: Orthoptera) are often stripped dark and pale green which make them hard to see when they are sitting on grass stems. Other grasshoppers are green and are difficult to distinguish from leaves e.g. long horned grasshoppers (Fig. 17) (Orthoptera, Tettigoniidae), still others have brown areas on their wings that resemble disease patterns commonly found on host leaves. Some grasshoppers have sand colour resembling the stone pieces (Fig. 18).

Further, objects to be copied must be inedible or at any rate not sought out as food by birds or other animals. Leaves, green and alive or withered and brown are often mimiced by butterflies and moths whose broad coloured wings afford ideal material for natural selection. As already described, many butterflies have an outline when sitting with closed wings that closely resembles a leaf e.g. the leaf butterfly, *Kallima* spp. (Fig. 1 A,B).

One of the most important reasons for the colouring of butterflies and moths is to camouflage them and the Lepidoptera have some of the most fascinating examples of camouflage in the animal kingdom. Quite a number of butterflies bear a strong resemblance to leaves and perhaps to the most effective of these is the leaf butterfly, *Kallima* spp. (*vide supra*) (Fig. 1 A,B). The wings are the exact shape of leaves and the marks on the underside closely resemble the midrib and veins and furthermore, no two butterflies are exactly of the same shade and colour as others. Some are marked with spots like leaf decay. The camouflaging behaviour of caterpillars is also very impressive.

The stick caterpillars are some of the most fascinating and are different from stick insects of the order Phasmida. They are capable of taking on the exact shade of colour of the twigs (Figs. 2, 3) that they resemble and caterpillars taken from the same brood but reared on different bushes will be of different colours. These caterpillars which are all in the family Geometridae, have evolved bud like projections, scars and other bark-like markings so that they are most difficult to be detcted specially as they rest at an angle to the twig, head outwards forming a natural looking `V' shape. Some moths which mimic pieces of broken twig will drop to the ground if disturbed and lie motionless, (Fig. 14B) hence well concealed on the litter of woodland floor.

A

B

Fig. 13. **A.** *Limenitis dudu* the pupa appears like shrivelled leaf, **B.** A Dried leaf insect.

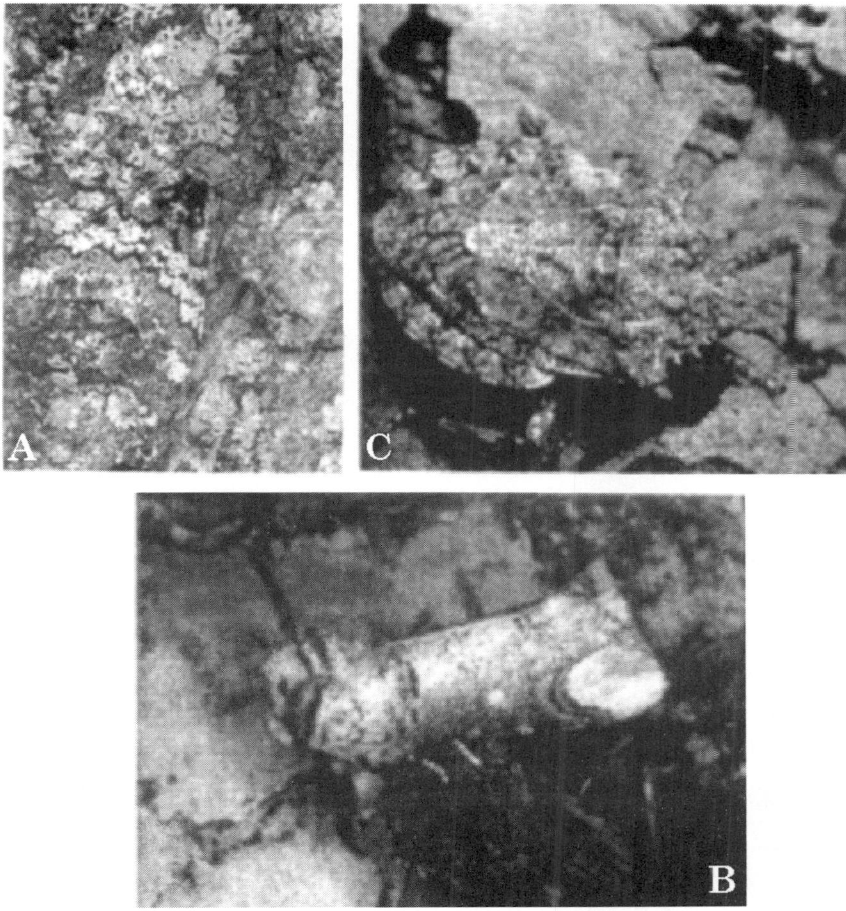

Fig. 14. **A.** A well camouflaged moth against moss and lichen covered bark. **B.** The moth looks like a broken twig, **C.** A stink bug seems like a piece of bark.

Several moths have a remarkably colour pattern on the wings like that of moss or lichen growing on a tree trunk (Fig. 14 A), so that they virtually disappear when at rest on this type of background. Similarly the young caterpillars of lemon butterfly, *Papilio demoleus* (Papilionidae) is to imitate the droppings of birds. Likewise, stink bug appears like a piece of bark (Fig. 14 C) The behaviour of insects usually plays an important role in making its camouflaging colour and pattern effective. Every insect must choose a suitable background on which to rest and furthermore, when at rest it must keep perfectly still.

Fig. 15. A leaf insect

A B

Fig. 16. **A.** Monarch *Danaus plexippus* (distasteful model), **B.** Viceroy, *Limenitis archippus* (Mimic)

In addition to the morphological adaptations, some insects conceal themselves with available material from the surroundings. Certain leaf beetles become less conspicuous. The larvae of caddisflies (Trichoptera) use a variety of materials of their surroundings e.g. pebbles, gastropod shells and twigs etc. to form cases that blend with the background and also provide physical protection (Fig. 19 B-E).

Fig. 17. A long horned grasshopper resembling green leaves.

Fig. 18. A ground grasshopper appears like stone.

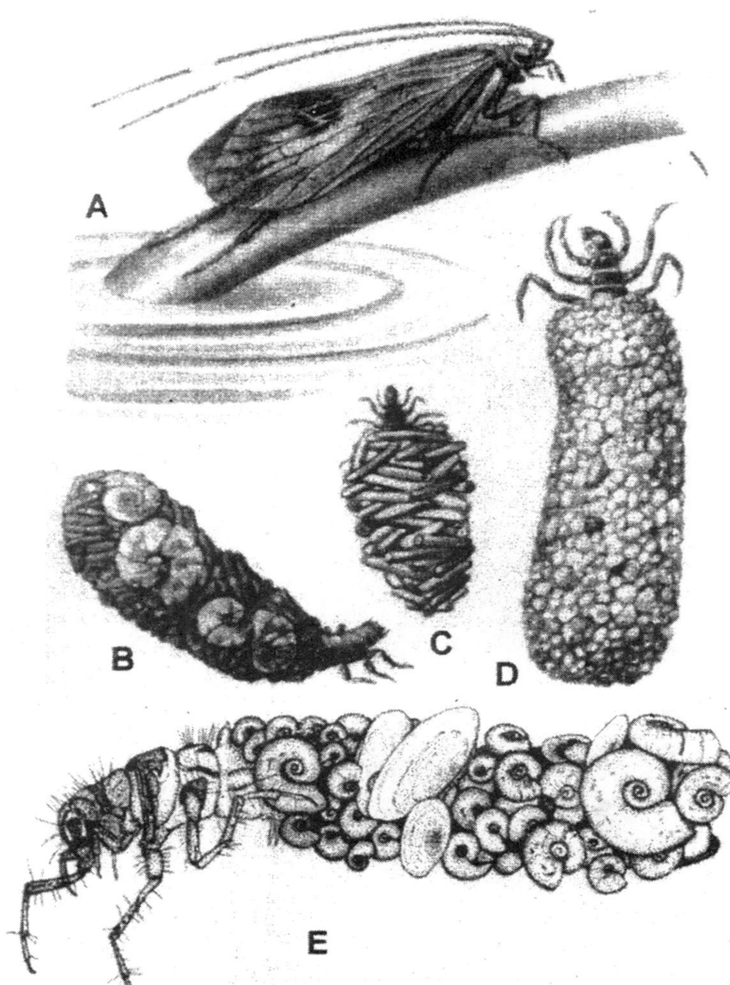

Fig. 19. **A.** Adult, **B.-E.** Larvae of Caddis fly, note nest builing material

A number of insects are noticeable by the fact that their colours seem designed to make them conspicuous against the background of their habitats. Many of them are boldly marked with black and yellow or black and red. Smelling or tasting these insects will often reveal an unpleasant odour or flavour and some of them such as wasps are armed with venomous stings as pointed out earlier.

Young birds do not have an instinctive fear or dislike of these insects but one or two trials, in which the insect may be fatally injured are enough to teach the bird that these are creatures best left alone. The more conspicuously coloured the insect is the more readily will be the birds learn their lesson and for the sacrifice of a few of their number the rest gain a valuable degree of immunity from attack.

Among the commonest of insects protected in this way are some of the bugs of the order Heteroptera which are often black with startling red, yellow or white patterns.

DEFENSIVE BEHAVIOUR

Skeletal structures, although not always used for defence, may function in this capacity. The raptorial forelegs of a praying mantis (Fig. 20A) although functioning primarily in prey capture, are also occasionally used to strike out at an attacker. Likewise, the legs of some bush crickets are provided with sharp and pointed spines which are used for defence. These are capable of piercing the flesh of a bird or mammal (Fig. 20B). The stings of a large number of ants, bees and wasps also serve such a double role.

Fig. 20A. The praying mantis (*Mantis religiosa*)

Fig. 20B. The spiny legs of this bush cricket are used for defence.

Further, the bites of many insects are accompanied by the injection of poison produced in special glands associated with the mouthparts. Similar glands associated with the egg-laying apparatus occur in certain bees and wasps. Here the ovipositor has become modified into a sharp, pointed sting (Fig. 20C) specifically for the injection of venom and eggs are laid without its aid. The venom of these insects produces a very painful reaction and can be dangerous to humans.

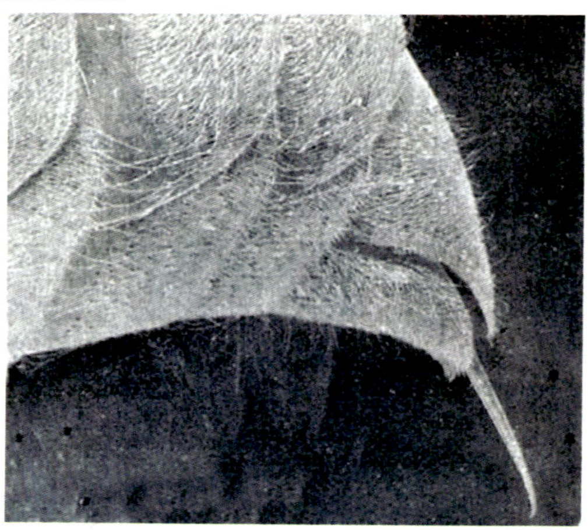

Fig. 20C.Abdomen of a vespid wasp (*Polistes* sp.) with the terminal sclerites parted and sting thrust into stinging position as seen through the scanning electron microscope.

The sound producing locusts, grasshoppers and crickets produce sounds of warning and aggression. Any skeletal structure of sufficient strength and hardness can serve in a passive defence capacity. The cuckoo wasps (Hymenoptera: Chrysididae) roll up into a ball when threatened. In this position the exposed areas of exoskeleton apparently provide protection from their hymenopteran host.

Numerous insects produce non-skeletal structures of different types in the form of nests, cocoons, cases etc. which function in defence. The predaceous larvae of the green lacewing, *Chrysopa slossonae* (Neuroptera) feed on the woolly alder aphid, *Prociphilus tesselatus*. The lacewing plucks the waxy wool from the bodies of the aphids and applies this material to its own body, thus coming to look like its prey. This disguise protects it from the ants that accompany the aphids.

The principal defence mechanism of insects is chemical which is secreted by repugnatorial glands. Repugnatorial material is discharged from storage organs that are associated with the glands by a variety of ways e.g. the secretion oozes on the integumetal surface. The gland is evaginated and the secretion allowed to volatilize or the secretion is forcibly discharged.

The caterpillars of the species, *Papilio* (Lepidoptera : Papilionidae) possess glands behind the head, known as *osmeteria* (Fig. 20D) which they evert when threatned that produce a secretion effective against ants. The forcible discharge of material takes place in many ways. The insects that are capable of forcible discharge of odoriferous material are stink bugs, caterpillars and carabid as well as tenebrionid beetles. Some of these insects are able to spray several metres. The peculier and impressive discharge mechanism is found in the bombordier beetles, which eject a hot spray containing quinones from the glands located in the posterior part of the abdomen with a distinct explosion. This is due to the result of a chemical reaction in which oxygen is liberated. It is very interesting to note that why the defensive secretions liberated by insects do not harm the insects themselves. This is because the gland cells allow two or more harmless precursors to be mixed in sacs or chambers isolated from the rest of the cell. This principle works well at the multicellular level in the repugnatorial glands of bombordier beetles. This can be explained in the following way-each gland which is two chambered contains phenolic precursors of quinones and hydrogen peroxide in an inner chamber and the enzyme catalase in the outer chamber. When these substances come into contact with one another the oxygen is liberated as a result of the mixing of catalase and hydrogen peroxide, and the phenols are oxidized to quinones, producing the reaction as described above.

Fig. 20 D. A caterpillar of *Papilio* sp. showing defensive attitude.

It may be pointed out here that besides functioning as defence allomones (the chemical emitted by an individual of one species and received by another species that induces a response favourable to emitter), some of these chemicals have antibacterial properties. Some acts as alarm pheromones, as e.g. found in ants, bees and wasps. The products of venom glands associated with hymenopteran stingers and with the stinging hairs of many lepidopteran caterpillars (Fig. 8-10 A,B) may be placed alongwith the chemical defence class.

Further, it must be noted that all defensive chemicals are not directly associated with integumental glands. Some are found in the haemolymph and impart a bad taste or render an insect indigestible e.g. the Monarch butterfly, *Danaus pelxippus* possesses such substances called cardenolides in its body tissues. Cardenolides are derived from the food plants, the milkweed *Calotropis* species.

As remarked earlier the Viceroy butterfly, *Limenitis archippus* is a well known mimic of monarch butterfly and is avoided by experienced vertebrate predators, even though it does not possess any distasteful substances and is quite edible. In a similar manner some other insects that feed on milkweed plants e.g. *Limnas chrysippus, Limnas alcippus* (Lepidoptera : Danaidae) and the milkweed grasshopper, *Poecilocerus pictus* also contain distasteful substances and are avoided by birds.

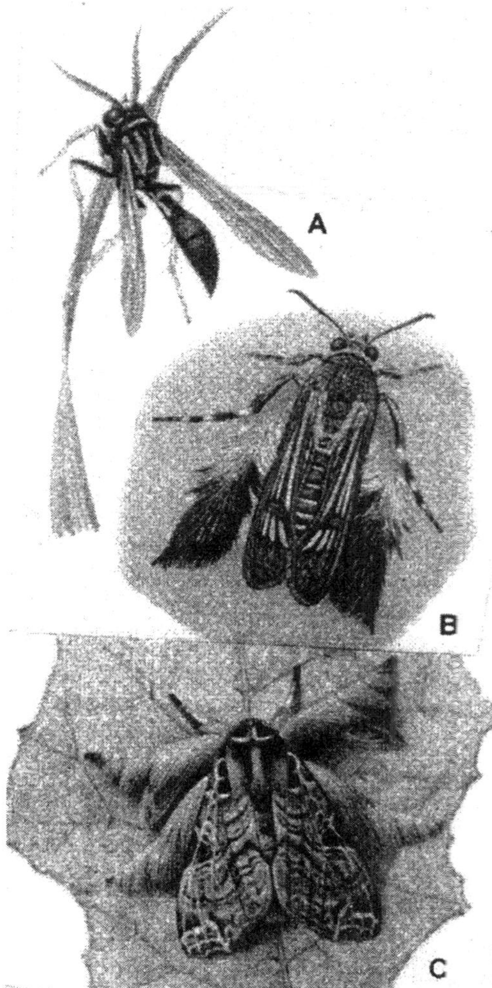

Fig. 21. A. A harmless moth resembling a wasp. B. A clear wing moth looks like a dangerous hornet, C. Owlet moth mimicking a dangerous silk worm

The Monarch caterpillar feeds on species of *Asclepias* or *Calotropis* spp. milkweeds (Asclepiadaceae) and has the ability to accept their poisonous sap (i.e. cardenolides) into its body system without any ill effect and also to store it so that it is still present in the body of pupa and butterfly. It is not ill tasting but has the effect on birds of making them vomit violetntly and painfully very soon after their ill chosen meal, and they associate their meal with the discomfort which they have suffered.

The milkweed bugs, like *Oncopeltus fasciatus* and *Lygaeus militaris* (Lygaeidae : Heteroptera) contain cardenolides from their food. Most of the insects that produce repugnatorial substance, taste bad or sting, are brightly coloured. This aposematic or warning colouration advertises their undesirability as food.

3

SOCIAL BEHAVIOUR AND PARENTAL CARE

The social insects are those wherein the parent and offspring live together and lead their lives in mutual cooperation under a common shelter. Such a condition is possible only when the length of life of female is prolonged with their children. Such social insects are earwigs, termites, certain beetles, bees and some wasps. The truly social insects belong to order Isoptera and Hymenoptera.

CLASSIFICATION OF SOCIAL INSECTS

The classification of social behaviour is based on the degree of involvement (see: Michener, 1969; Horn, 1978; Gullan and Cranston, 1981 for detail) and they have been broadly classified as follows:

Sub-Social Insects: Sub-social insects are those wherein adult care for young. A range of sub-social behaviour is seen in a few species of thrips (Thysanoptera), aphids (Homoptera) and some Hymenoptera.

Communal Insects: Communal insects are those in which the members of the same generation use a composite nest when favourable conditions for nest construction are scarce and scattered throughout the environment, communal nesting may occur. Communal nesting may stem from daughters choosing to nest in the natal nest giving enhanced utilization of nesting resources and mutual defence against parasites.

Quasi Social Insects: It is of interest to note that in quasi social behaviour, a communal nest consists of members of the same generation, all of which assist in brood rearing and all females are able to lay eggs even if not necessarily at the same time. Thus in short quasi social is like communal but in which adults cooperate in brood care.

Semi Social Insects : In semi social behaviour the communal nest similarly contains members of the same generation cooperating in brood care but there is division of reproductive labour with some females (i.e. queens), laying eggs whereas their sisters act as workers and rarely lay eggs. The semi social behaviour differs from eusocial behaviour only in that the workers are sisters to the egg laying queens rather than daughters as in the case of eusocial society and as in primitive eusocial hymenopterans there is no morphological (i.e. shape and size) difference between queens and workers. Thus, the semi social insects are like quasi social but here workers care for young of reproductive.

Eusocial Insects : In which truly social insects are found in overlapping generations and offspring assist their parents in maintenance of the colony e.g. Isoptera and Hymenoptera (Figs. 22-24 A,B). The eusocial insects have a division of labour in their colonies with a caste system comprising a restricted reproductive group of one or several queens, aided by workers i.e. the sterile individuals that assist the reproducers and in termites and many ants, an additional defensive soldier group. There may be further division into subcastes that perform specific tasks. At their most specialised members of some castes such as queens and soldiers may lack the ability to feed themselves. The tasks of workers, therefore, include bringing food, to these individuals as well as to the brood i.e. the developing offspring.

Fig. 22. A colony of termite

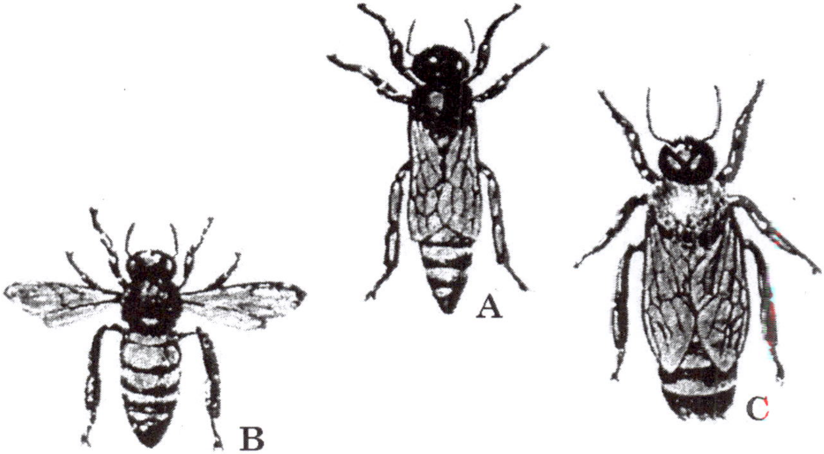

Fig. 23. A. Queen or female honey bee, B. Worker or unfertilized female honey bee, C. Drone or male honey bee

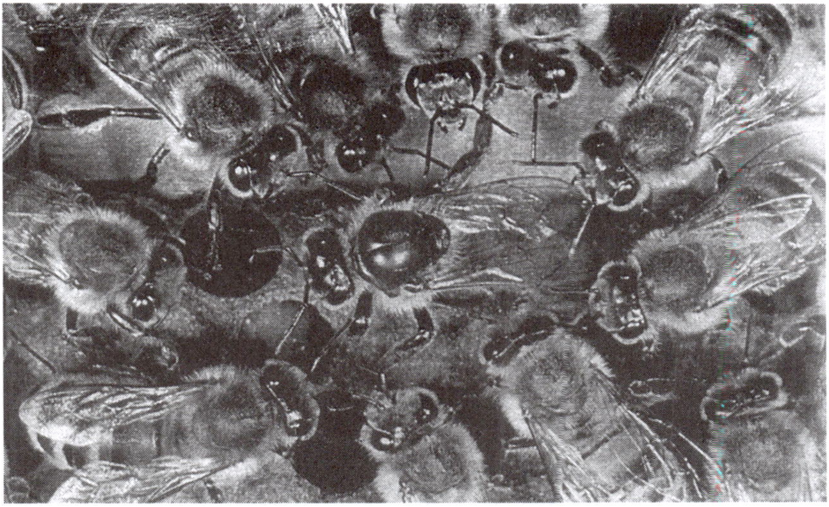

Fig. 24A. The queen honey bee in the centre surrounded by workers.

The primary differentiation is female from male. In eusocial Hymenoptera which are haplodiploid in sex determination, queens control the sex of their offspring. Releasing stored sperms fertilizes haploid eggs which develop into diploid female offspring whereas unfertilized eggs produce male offspring. At most times of the year queens are rare compared with sterile female workers. Males do not form castes and may be short lived, dying soon after mating. In termites, (Isoptera) males and females may be equally represented with both sexes contributing to the worker caste. A single male termite i.e. the king may permanently attend the queen.

ISOPTERA

Termites

All the termites may be divided into two groups - the lower and higher termites. The species - rich higher termintes (Termitidae) differ from lower termites in the following manners :

A. The termitidae digest cellulose, using either predigestion by cultivated fungus, incorporation of the enzyme cellulase in the diet or the use of anaerobic bacteria for internal digestion. This contrasts with the dominance of symbiotic internal flagellates in lower termites.

B. Termitidae have a more elaborate and rigid caste system e.g. in most lower termites there is little or, no distension of the queen's abdomen whereas termitid queens undergo extraoridinary physogastry in which the abdomen is distended to 500-1000% of its original size.

The members of this group form highly organized colonies called termitaria (Fig. 22). The colonies are polymorphic and five chief castes are recognized. In all these castes both sexes of individuals are present. Of these five, three kinds of functional reproductive forms and two kinds of sterile forms are present. Amongst the functional reproductive forms (male and female) the usual reproductive caste is the fully winged or macropterous form. The individuals of this caste immediately after emergence start mating and then loose their wings. Supplementary reproductive forms are brachypterous (neotenic) and apterous. In highly specialised termite colonies only a single fertile female is present. This is derived from any of the three above mentioned reproductive castes. The sterile castes are the workers and soldiers, the former being more numerous in a colony. Their chief function is to build the nest and supply food. The soldiers act as defenders and are morphologically differentiated from the workers in the formation of well developed mandibles.

Interestingly enough, Horn (1978) remarked that termites are unique among insect orders in that all species are eusocial. Typically a mated pair initiates the colony. After a short reproductive flight they shed their wings and seek out a favourable undergound area. Here the female lays eggs that hatch into tiny, blind nymphs which are the first workers. As these grow they take up the tasks of foraging, feeding the reproductives and the younger nymphs. All termites feed on cellulose containing foods. Some termintes feed on fungi which they grow in their gardens within the colony. Cellulose is digested by symbiotic protozoa, present in the hind gut, since termites eat one another's cast skins and faeces, thus conserving hind gut protozoa. In addition they also eat eggs and dead termites and thus get protein which is not present in cellulose.

HYMENOPTERA

Social wasps: Like Isoptera, social behaviour in this group exhibits caste differentiation only in certain forms. It may be pointed out here that the workers are sterile females developed from the fertilized eggs of the queen while the drones are males developed from virgin eggs parthenogenetically. The colony formation and the duration of a colony vary in different groups. Amongst the social wasps, the annual and perennial colonies are produced. The annual colonies are founded by the wasps of the temperate regions and a single fertilized female is the founder. Such colonies are termed ``monogynous" e.g. *Vespa germanica* and *Vespa sylvestris* (Hymenoptera : Vespidae). The tropical Vespidae form perennial colonies and they are "polygynous". Here the colony formation is by the combined effort of a number of females and workers.

Nest of a social wasp resembles a honey bee's in one respect i.e. it contains batteries of hexagonal cells (Fig. 25A) in which eggs are laid and larvae grow and pupate. Nothing could be more different however than the materials from which they are made. Bees eat quantities of pollen rich in protein and produce from it wax which they excrete, mould with their jaws and use to build their comb. The wasps on the other hand, scrap the surface of dead wood with their jaws and make a pulp with saliva which hardens to form a tough paper.

Thus, certain wasps prepare their nests either with cellulose i.e. paper Wasp (Fig. 25A) and/or others prepare beautiful mudnests of different types (Fig. 25 B,C) by clay.

Social bees : Among the bees, the group Bombidae or Bumble bees exhibits a remarkable social behaviour. Here the insects build their nest in some cavity in the ground and its walls are constructed of fragments of grass. The nest contains a lot of pollen mixed with honey. Over this provision, the female constructs the cell of wax and deposits her eggs. A waxen receptacle or honey pot is also prepared by the queen and this is filled with honey which acts as a food when the larvae hatch. They feed on this food and later on when this food is exhausted, the covering of wax is removed and the original provision i.e. mixture of pollen and honey is utilized. The workers hatched first when sufficient numbers are released. They allow the female for the sole purpose of egg laying. The difference in the development of queen and worker is based on the quantity of food rather than on quality. Both perennial and polygynous colonies are formed.

In Apidae, a honey bee colony (Figs. 23-24A) consists of several thousand individuals and the queen becomes highly specialized egg laying machine. The young larvae are first nourished on Royal jelly a product of pharyngial glands of workers. Feeding continues until pupation for would be queens (female reproductives), drones (reproductive males) and workers (non-reproductive female). It is replaced from the fourth day by a mixture of honey and predigested pollen.

It is interesting to note that the emergence of a queen, inhibits production of additional queen cells and now comes the moment for which the drones have awaited. The queen departs on a mating (nuptial) flight, the drones chase after her attracted by a pheromone and several males mate with her. Thereafter, she returns to the hive, to live out her life for about six or seven years. She does not mate again but fertilizes her eggs with stored sperms. Her major functions thereafter are laying eggs one every 30 seconds and secreting pheromones. Workers feed and groom both queen and drones, although the drones are forcibly removed from the hive when cold weather ends the honey flow. Unlike queen, the workers may live only a month during honey flow.

The worker commences as a house cleaning and nurse bee, cleaning out old brood cells, removing bee droppings and feeding developing larvae with pollen and nectar brought in by the foraging field bees. They also participate in the comb construction or guard duty at the entrance of hive. Hive bees also regulate temperature within the hive. Thereafter, hive bee progresses to being a field worker and makes several trips a day to collect pollen and nectar. If the worker bee survives predation, bad weather and several other abnormal conditions, then she spends her old age as a scout bee, exploring for new food sources at great distances from the hive and if successful returning to do the wagging (waggle) dance.

Fig. 24B. Honey bees swarm

Further, it is interesting to mention that if a honey bee colony becomes overcrowded it usually sends out a swarm, consisting of the old queen and a few thousand workers - usually about half of the original population. The swarm usually settles on a tree or a bush (Fig. 24B) and may stay for several days while scout bees look for a permanent home. If they find something promising, they return to the swarm and `dance' on its surface. The dance indicates the direction and also the distance of the site, and it encourages other workers to go and have a look. If enough workers come back and indicate a particular site, the whole swarm moves to it and the workers start building new combs.

It is well known that honey bees are able to regulate the temperature of their hive. When the colony gets too hot, the bees cool it by bringing water, which is deposited on the upper parts of the combs. Then they fan vigorously with their wings and drive the air, cooled by evaporation, through the brood comb. On the other hand in winter, the workers form a dense cluster which reduces heat loss and they can warm the hive if necessary by running about. This increases the rate of metabolism so that their bodies are warmed. To a lesser extent, bumble bees are also able to control climatic rather microclimatic conditions within their nests (Cloudsley-Thompson, 1967).

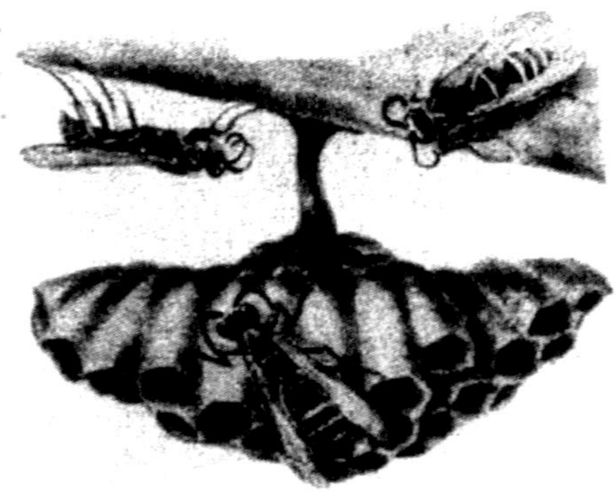

A

B

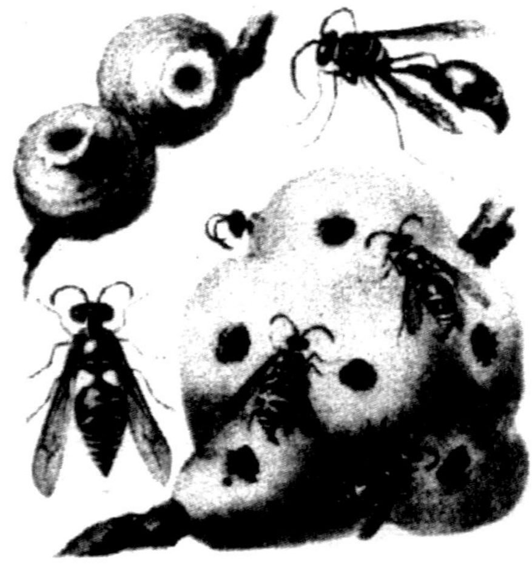

C

Fig. 25. A. Paper wasps and nest.
B-C. Mud wasps. B-Potter wasp and nest. C. Mason wasps and nest.

Social ants : All ant socieities are divided into three castes viz. queens, that form new colonies and thereafter function as egg laying machines; winged males that take a nuptial flight, once with a queen, fertilizing her for life and thereafter die and sterile females that lead tirelessly neuter lives work hard at an amazing variety of skill (routine work) needed by the community. Some act as nurses, some as house keepers, some as hunters or even soldiers.

The nests of ants though enormously populous, show a little sign of planning or design. They do not rear their young in cells but carry them from one part of the nest to another and the nest is usually simply an intricate arrangement of tunnels and chambers hollowed out of earth. Ants do the most extraordinary things. They keep other insects as domestic animals and other kinds of ants as slaves. They practice precise and elaborate forms of agriculture and make war on each other. They use their own larvae as instruments for spinning silk.

PARENTAL CARE

Parential care may be considered to be a social behaviour. The female insects select an appropriate oviposition site affording protection to the eggs and ensuring an appropriate food resource for the hatching offspring. The ovipositing female may protect the eggs in an ootheca, or deposit them directly into suitable substrate with her ovipositor or modify the enviroment as in nest construction. Parental care is seen as post oviposition and post hatching attention involving the provision and protection of food resources for the young. Although egg and early instar attendance is predominently a role of female individual yet parental guarding is seen in some Hemiptera e.g. giant water bug (*Belostoma* : Belostomatidae) and tree hoppers (Membracidae) (Figs. 26, 27). In *Belostoma* the female bug oviposits on to the dorsum of the male which receives eggs in small batches after each copulation and each male *Belostoma* may even accept eggs from more than one female. Other functions of parental care include keeping the eggs free from fungi, maintaining appropriate conditions for egg development, herding the young and feeding them. Likewise, nesting is also a social behaviour in which the parents use a structure in which to lay their eggs and to which they bring food supplies for the young. Nesting is seen in the order Dermaptera, Coleoptera, sub social Orthoptera and subsocial Hymenoptera. Earwigs of both the sexes over winter in a nest. In spring, the male is ejected when the mother starts to tend the eggs

(Fig. 28). In some species the mother earwigs forage and provide food for the young nymphs. Mole crickets, *Gryllotalpa* spp., exhibits similar behaviour. Some scarabs roll the dung away from its source, others coat the dung with clay. Both sexes cooperate but the female is mostly responsible for burrowing and preparation of the larval food source. Eggs are laid on the burried dung and in some species no further interest is taken. In others, parental care is well developed, commonly with maternal attention to fungus reduction, and parental defence to remove or exclude ants.

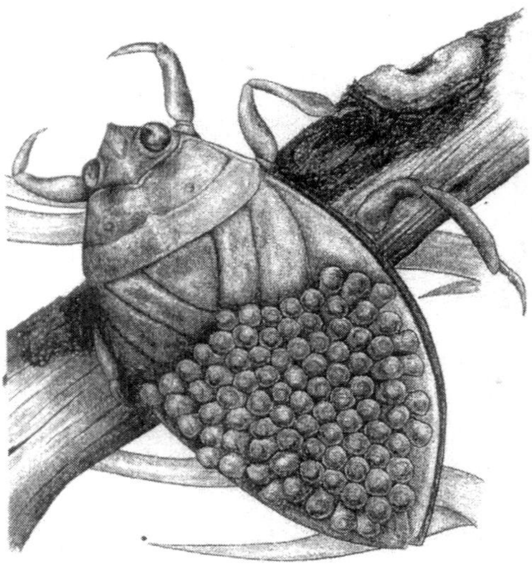

Fig. 26. A giant water bug, male with eggs on the back

Fig. 27. Female tree hoppers brooding nymphs on a twig

Amongst the Hymenoptera, subsocial nesting is restricted to some aculeate Apocrita viz-within the super families Chrysidoidea, Vespoidea and Apoidea. In bees and masarine wasps, pollen replaces arthropod prey as the food source collected and stored for larvae. Nest complexity in aculeates ranges from a single burrow provisioned with one food item for one developing egg to the linear or radial multicellular nests. Several types of construction media are used like soil, stems, rotten wood, solid wood as in carpenter bees (Xylocopinae).

Thus, we find that social behaviour in insects is a remarkable feature exhibiting a wonderful coexistence amongst the individuals of the group forming a colony. It has reached its highest peak only in the case of Isoptera and Hymenoptera. The significance of social behaviour and the chief principle underlying this phenomenon is the division of labour.

Fig. 28. Earwig protecting her eggs as well as newly hatched youngs

ADVANTAGES AND DISADVANTAGES OF SOCIALITY

In previous pages a brief account of social organization of some insects has been given. In the following pages advantages as well as disadvantages of sociality have been discussed. Wilson (1972, 1975) has remarked that although there are many advantages to sociality yet there are disadvantages as well. And among the `antisocial' aspects of phylogenetic inertia are outlined as the following.

1. *Limitation of food* : A large concentration of food is necessary to support a large group.
2. *Reproductivity effect* : Animals living in large groups tend to produce fewer offspring per animal than those living in smaller groups.
3. *Sexual selection*: When sexual selection has favoured development of large, aggressive males, they are less likely to be integrated into organized societies.
4. *Danger of inbreeding*: Where a tightly knit social structure limits breeding choices, inbreeding can occur and thus reduce individual fitness levels.
5. *Exposure to disease and predators :* A grouping of animals increases the chances of spread of diseases and under some conditions may increase the risk of detection by predators.

Wilson (1975) and Klopfer (1974) have further pointed out that since social organization has evolved repeatedly there are obviously many factors that counteract the antisocial forces and act to produce the evolution of organized societies. Among the factors working to increase sociality are :

1. *Defence against predators* : While groups of insects may be more likely than isolates to be detected by predators, they also have improved defence through group action.
2. *Increased competitive ability* : Just as groups are able to obstruct predators, they can overwhelm animals of another species with which they compete for identical resources.
3. *Buffers against environmental processes* : Animals in groups buffer each other from the environment as in the protection against wind and the dilution of environmental toxins.
4. *Increased feeding efficiency* : Feeding efficiency may be increased through either imitation of foraging habits or cooperation in the procurement of food.
5. *Penetration of new adaptive zones* : Grouping may on occasion permit animals to enter novel habitats or to assume a new life style.
6. *Increased reproductive efficiency* : Animals in groups can more easily locate mates and synchronize breeding.
7. *Increased survival at birth* : Groups of animals are often more able to ensure survival of offspring than are isolates.
8. *Improved population stability* : Social living may increase individual fitness levels by ensuring that population size remains within tolerable levels appropriate for the habitat.
9. *Modification of the environment* : Large groups can alter the environment - as in building structures (termites) regulating environmental temperature (honeybees) or altering plant growth - as the result of their presence and behaviour.

4

THE LANGUAGE OF BEES AND THEIR BEHAVIOUR

The honey bees are one of the most intelligent socially well organized and remarkable amongst insects. Their brain (memory) and olfactory sense organs are very well developed. Honesty, sincerity, dedication, division of labour etc. seems to be the guiding principles of the bees-society. The various castes perform their duty with full responsibility.

The life of bees is extremely complicated and mysterious but at the same time very systematic.

No insect group has such a wonderful, well disciplined society in all the sense as the honey bees have. One of the acts of the bees is their communication system, which is truly perfect and is largely based on various types of dances and the specific odours the pheromones.

In the following pages, only one aspect i.e. how the bees talk and convey their message to other bees is briefly described on the basis of the work of Karl Von Frisch, who dedicated his entire life on the study of honey bees and thier behaviour with meticulous care. Frisch remarked that "the language of bees is truly perfect and their method of indicating the direction of food sources is one of the most remarkable mysteries of their complex social organisation".

THE DANCING BEES

Every one is very well aware of the dancing and humming of bees on flowers. Their different methods of dancing have different meanings and are not a useless part of their daily life. It is by means of these methods that the bees talk and by which they tell other bees about the distance and direction of the food source. Besides the methods of dancing, there are other ways also by which they tell their friends about flowers full of scented nectar and ask them to go there for collecting food. Inspite of being small cretures they show a remarkable intelligence and this is one of the wonders of nature.

How the bees talk and how they tell their mates about the presence of food at a new location is quite interesting. When a dancer bee, during its search for food, comes near an interesting flower, it dances around it for a while, collects some scented nectar from it and then returns back to its honey comb, where it dances again, telling others about the flower. During the dance, the bees nearest to the dancer become very much excited and soon join the dance, keeping their antennae close to their bodies. After a while, the dancer turns away and leaves the hive others follow and soon a small procession reaches the feeding ground, the process being repeated many times. Thus it is clear that it is by this dancing, that one bee convey the information about explorable feeding grounds to other bees.

The question now arises as to how the dance tells about the distance and direction of food source.

Round Dance

When a dancer bee dances in simple circles (Fig. 29A) it means according to Frisch, fly out and seek in the neighbourhood of the hive. This round dance indicates food in near vicinity i.e. less than 100 metres of the hive. The surface of a bee's body has the ability to hold scents for long periods and the dancer also regurgitates droplets of nectar it collected. Other bees perceive the scent on the dancer's body surface and that present in the regurgitated nectar. They start in the direction indicated by the dancer and soon search out the flower, with whose scent the two scents perceived by them tally.

It is noticeable that these dances are made only when there is a rich source of food. According to Frisch these dances also carry the basic message. There is plenty of food and sweetness. If there is no food, then the dancer bee, after returning to the hive, does not dance.

When there are a large number of flowers rich with nectar, the dancer collects nectar from the richest flower, returns to its hive and dances vigorously; sending a large number of bees to that particular flower. When, after repeated attacks, the nectar supply of flower gets exhausted, the returning bees no longer dance on reaching their hives; with the result that no more worker bees go out to the already exhausted food source. In case nectar of the flower concerned is not much scented, bees make use of the scent gland found on their abdomens. It secretes a scented fluid, which the bees pour on the nectar and other bees locate the flower by this particular scent, they are so familiar with.

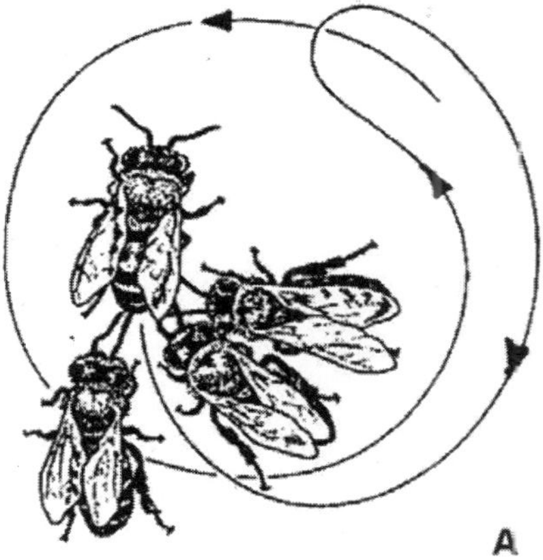

A

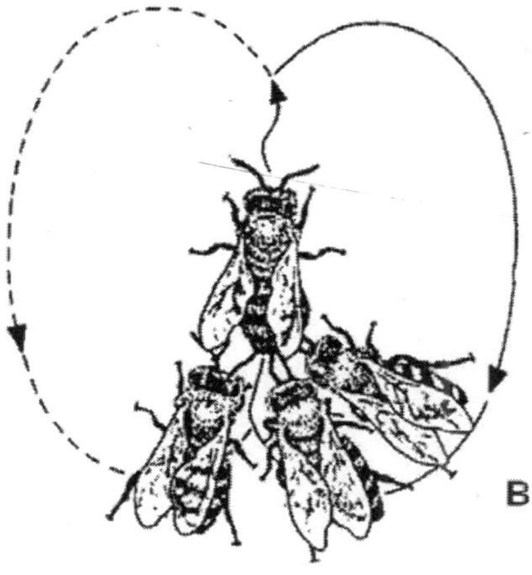

B

Fig. 29. A. The round dance of honeybees. The round dance is given when a food source is located less than 100 meters from the hive. B. The waggle dance of honeybees. The waggle dance is performed when food is located more than 100 meters from the hive. During the straight run in the dance, the bee's abdoman moves from side to side and produces a buzzing sound.

Waggle Dance

When food source is at a very distant place, i.e. more than 100 metres, the dancer bee which comes from that feeding place, dances in a quite different manner. This particular type of dance, Von Frisch termed as `wagging or waggle dance' (Fig. 29 B). The dancer runs a short distance in a straight line, wagging its abdomen very rapidly from side to side; turns to the left; and makes complete angle of 180°. This process is repeated many times.

Thus, as observed by Frisch, there are two types of dances; the round dance indicating presence of food source in near vicinity of the hive and the wagging or waggle dance which means that the food is at some distance sometimes many kilometres away from the beehive. As a matter of fact the wagging dance not only announces a rich source of food at a far off distance more than 100 metres but also tells how far away and in what direction. The actual distance and direction is indicated by the number of turns made during the dance (Fig. 30 A-D).

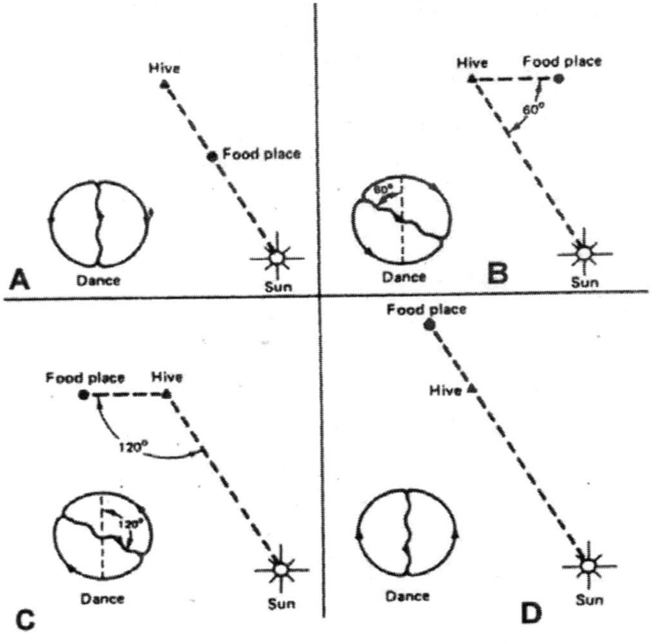

Fig. 30. The dance language of the honey bee. Four examples of indication of direction. The sun is to the south-east of the hive. The diagram in each lower left corner shows schematically the direction of the waggle dance for a given position of the feeding place.

Briefly speaking, when a rich source of food is found, its location is signalled to other forager bees and the returning dancer bee performs a ritualized dance on the vertical comb of the hive. The dance attended by other forager bees who are able to locate the food source on the basis of information provided by dancer. If the food source is less than 100 metres from hive the round dance is performed and other bees leave the hive and locate the food. If the food source is more than 100 metres from the hive, the dancer bee performs a wagging or waggle dance (Fig. 30 A-D) that contains information about the distance and direction of food, both. The distance is indicated by the duration of the straight run of the dance. Direction is indicated by transforming the direction to the food source relative to the location of the sun into an angle of the straight run of the dance relative to gravity (Fig. 30C). The other forager bees leave the hive and locate the food source and collect the nectar.

Further, it may be added that the Waggle i.e. the figure of eight (8) dance is also used when bees are searching for a new home. In a flourishing hive, part of the colony may emerge with the old queen as a swarm and establish themselves in a new locality (as pointed out earlier in chapter 3). First they cluster on any suitable object such as a branch of a tree. Then scout bees are sent out to find a permanent home. Each successful scout announces the locality of her favoured site by means of the dance performed on the surface of the cluster of bees. Other bees are recruited by the dancers to inspect the various sites offered and the suitability of each site is reflected in the vigour and enthusiasm with which returning bees dance. Eventually, one of the potential sites gains most favour and the swarm `agrees' on this before moving off. This process may last for several days, and on one famous occasion Martin Lindauer watched a swarm for four days, noting directions of potential sites indicated by the dances and calculating their distances. Eventually, with the aid of stopwatch and compass, he had the locality of the site which was rapidly gaining favour and, armed with this knowledge, he was able not simply to follow the swarm to its new dwelling but to arrive there before the bees did! (see Blaney, 1977).

Thus these small creatures talk to each other in their own language. The capability of talking with each other, though in a much different language of their own is a great asset to those insects which have chosen such a high grade of social organization. That tiny animals like bees should have such remarkable capabilities, indeed makes one marvel at the wonders of nature.

5

INTERACTIONS BETWEEN INSECTS AND PLANTS AS RELATED TO POLLINATION

It is well known that the sexual reproduction in the flowering plants is accomplished by the transfer of pollen from the anthers of a male flower to the stigma of a female flower. The transfer of pollen from a male to a female flower is accomplished primarily by the wind or by the activities of insects that associate with plants e.g. wind pollinated plants (anemophilous) include cereal plants like wheat and corn and many species of trees. The flowers of these plants are generally small with weakly developed petals, that do not produce nectar but produce dry pollen grains which are easily picked up by the wind.

Most flowering plants are pollinated by animals especially insects and it is argued that the success of the angiosperms relates to the development of these interactions.

Further, the benefits of insect pollination over wind pollination include : increase in pollination efficiency including reduction of pollen wastage, successful pollination under conditions unsuitable for wind pollination and maximization of the number of plant species in a given area, since even rare plants can receive conspecific pollen carried into the area by insects. (Gullan and Cranston, 1994).

RELATIONSHIPS BETWEEN PLANTS AND INSECT POLLINATORS

The relationships between plants and their insect pollinators are mutualistic in that the plants provide sugar rich nectar and/or pollen for insects, which in turn serve as accessory plant `sex organs' in transmitting gametes from the male to the female plant (or flower). According to Baker and Baker (1973) some insects obtain aminoacids from nectar and a distinct correlation is noted between the aminoacid concentrations in nectar and the dietary requirements of pollinating insects. Butterfly pollinated flowers averaged a greater aminoacid concentration than bee or bird pollinated flowers, as birds and bees have other portein sources (e.g. insects and pollen). Further, syncrhonous flowering often occurs in response to photoperiod and thus assures that there will be as many flowers as possible available for pollination. Baker and Baker (1975) have further confirmed that nectar of certain plants has been found to contain nutritive substances like aminoacids and lipids in addition to sugar. Insect pollinated plants have evolved traits that make them attractive to effective insect pollinators. In many cases the result has been a one to one relationship between a single plant species and a single insect species e.g. the yucca plant and yucca moths (*Tegeticula yuccasella*). An adult female moth uses its specialized mouth parts to gather the pollenia of the yucca plant into a ball. This ball is then scraped across the stigma of plant accomplishing cross pollination.

The dependence of the yucca plant on the pollinating activity of the moth is demonstrated by the failure of the plant to develop pods in the absence of the moth. The yucca plant provides a home for the moth larvae which remain within the plant until ready to emerge as adults.

Distinct correlations can be made between the anatomical and physiological characteristics of the flowers of a given species and the anatomy, physiology and behaviour of their insect pollinator. The characteristics of flowers attractive to insects include the production of particular scents, colour, size, and shape of petals and the shape of the flower.

The production of scent is very important in attracting a potential pollinator and often in determining behaviour following landing. Flowers pollinated by true flies which breed in dung may actually mimic the odour of dung, so ``fly" flowers commonly have a disagreeable odour for humans. On the other hand ``butterfly" and ``moth" flowers usually have a sweet smell. Flower colour may also be correlated with the type of pollinating insect e.g. honey bees and their relatives are attracted by many blue, purple and yellow flowers but cannot see red flowers. However, the butterflies are attracted to red flowers.

Some orchids e.g. *Ophrys* and others produce stimuli that attract their specific pollinators, particularly Diptera and

Hymenoptera, by releasing copulatory behaviour. These orchids produce scents that mimic the effects of sex pheromones as well as stimulating the insects visually and tactilely.

The most common insect pollinators are various members of the orders Coleoptera, Lepidoptera, Diptera, and Hymenoptera. Among the common adaptations of pollinating insects are elongated mouth parts (Figs. 35, 36AB) that facilitates obtainng nectar from flowers with deep nectaries, plumose hairs on the body to which pollen clings (Fig. 33) and various specialized pollen collecting and/or transporting structures like pollen baskets on the hind tibaiae of honey bees (Figs. 31, 32, 34A), and their relatives and pollen brushes on the hind legs of the most bees.

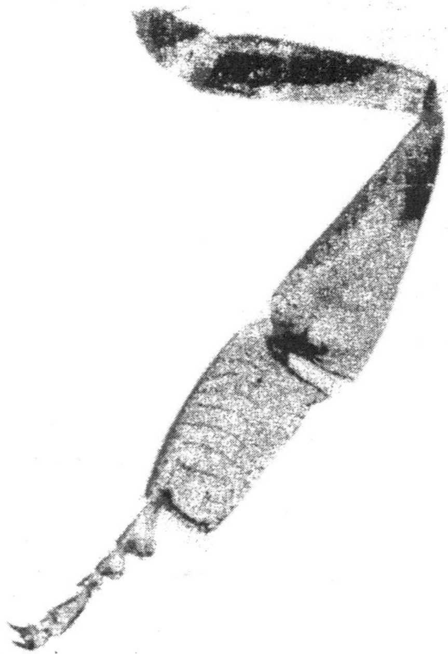

Fig. 31. Hind leg of a worker honey bee, showing pollen basket.

As already remarked, the main association of mutual benefit between insects and plants is the act of pollination. The landscape is full of flowers that must be pollinated by insects in order to produce fruits and seeds. The pollen must be transported from plant to plant so that each kind of flower receives its own kind of pollen. Furthermore, the pollen must be placed precisely on the female parts of the flower and this must be done swiftly, because many blossoms remain open for only a short time. A very large number of insects accomplishes this immense undertaking with a remarkable efficiency. The ants and beetles crawl over the flowers, butterflies and moths dip their long sucking mouth parts into blossoms, flies and wasps fly with buzzing or whizzing sound through the sea of petals and the efficient and purposeful bees tirelessly make their rounds. As already pointed out plants have developed attractions for insects in the form of colour, nectar and pollen. The insects brushes against the pollen bearing organs of the flower, becoming dusted with pollen (Fig. 33) which it carries to another flower of the same species. Worker honey bees possess dense body hairs that catch pollen (Figs. 33-34) which is removed by comb like structures on the legs and transferred to pollen baskets on the hind legs (Figs. 31, 32, 34 A). Many insects have greatly elongated siphoning type of mouth parts that can reach the nectar deep inside bell and trumpet like flowers.

Fig. 32. Honey bee with pollen basket filled with pollen grains

Interestingly enough, it has been commonly observed that the flowers display numerous other mechanisms for getting the full benefit out of search of insect for nectar and pollen e.g. some blossoms open and close only at certain times of the day to accomodate honey bees that have internal clocks to help them keep track of time and visit the right flowers at the right hours. The lady's slipper orchid has a nectar filled trap that imprisons the visiting insect : to escape, the visitor must take a path which not only pollinates the flower but dusts the insect with pollen for the next flower as well. Some orchids have evolved blossoms that look like female bees. The orchid puts out its flower as the male bees are emerging but before the females have left their pupal cases. A male, flying in search of females will eventually arrive at an orchid flower.

Fig. 33. The whole body of a bee is dusted with pollen grains

The mimicry of orchid flower is so realistic of the female bee that even the male is deceived. In its attempts to mate with the pollen bearing parts of the flower, it becomes dusted with pollen which it carries to the next orchid that seduces it. Although many kinds of bees visit flowers but none of them, not even the acclaimed bumble bee, possesses the remarkable efficiency of the honey bee at pollination. Honey bees are foremost among all insect pollinators because their whole economy is based solely upon the collection of nectar and pollen. (Figs. 32, 33, 34AB, 36 AB).

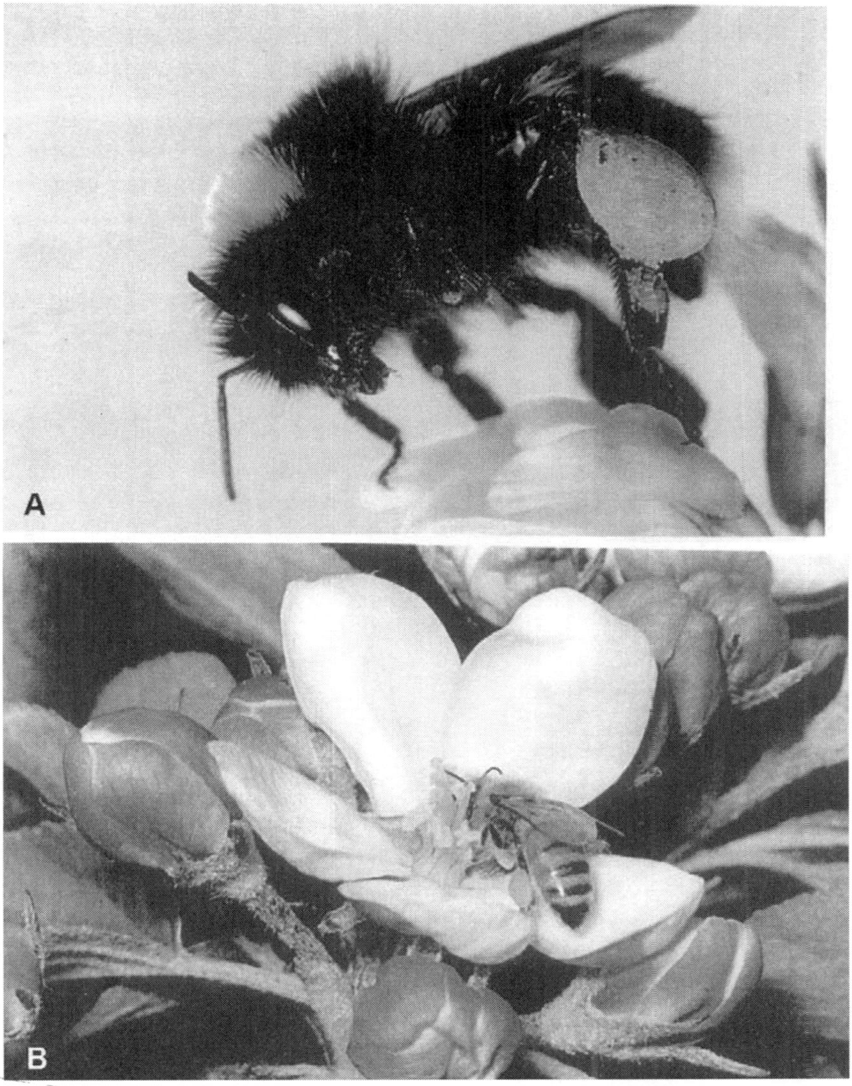

Fig. 34. A. Bumble bee's pollen basket filled with pollen grains B. A honey bee collecting nectar and pollen from apple flower.

It has been estimated that more than 65% of the flowering plants are pollinated by insects. There are major categories of plant adaptations that ensure or increase the efficiency of insect pollination are those that attract the insects to the flower so that pollen will be encountered and those that reward the insect with either pollen or nectar (*vide supra*).

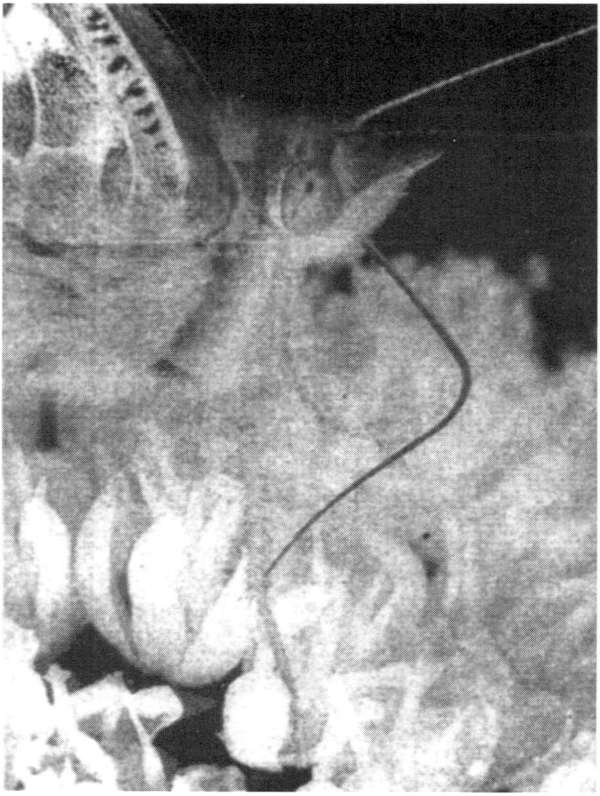

Fig. 35. A Butterfly, *Cynthia cardui* sipping nectar from a shallow bowl shaped flower of *Allium* sp. through its proboscis

Scientific experiments and practical experience have proved that cross pollination of flowers of entomophilous crops by insects is the most effective and cheapest method of increasing yields (see : Kozin, 1976). The cost of planned pollination of plants by bees is annually recovered 15 to 20 times over through the values of additional crop yield. It has been found that the increase in the fruit production in apple and pear is 50% to 60%, cucumber 75% to 90% and watermelon and muskmelon 95% to 100%. Even in tomato and grape which are primarily self pollinating, the fruit yield after cross pollination of flowers by bees increases 25% to 30%. According to results of studies carried out in different countries, the income from agriculture in regions of intensive cultivation obtained after using bees in plant pollination is 10 to 12 times greater than the cost of the honey and wax produced. Furthermore, it has been established that a large bee colony in the course of the year uses 100 kg honey

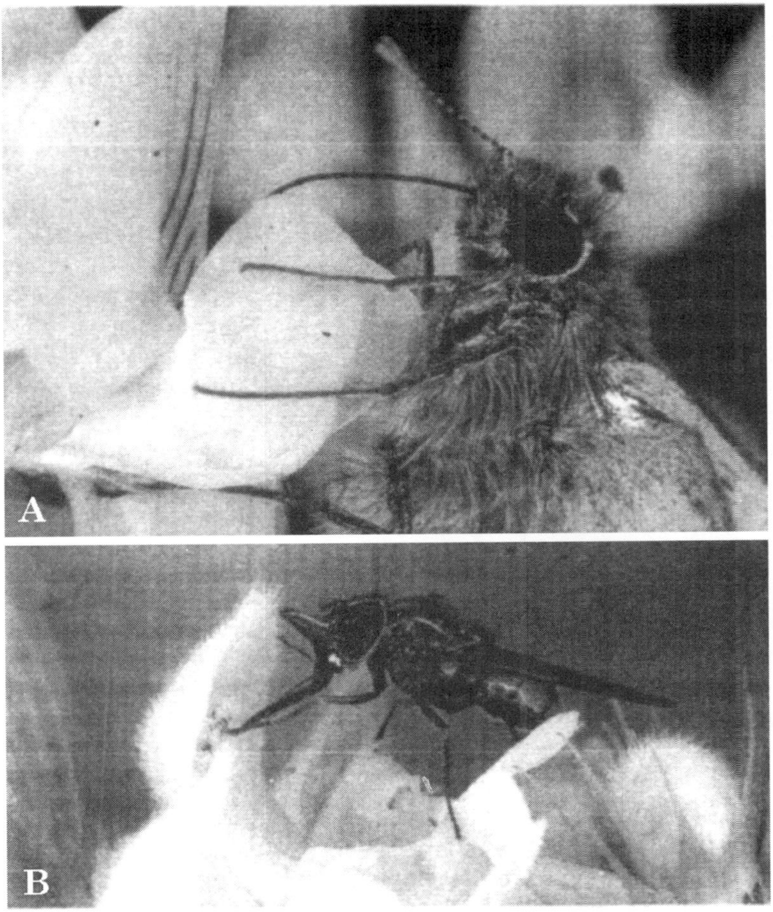

Fig. 36. A. A small skipper *Thymelious flavus* collecting nectar.
B. A Howerfly feeding on anthers.

and 30 kg pollen for feeding its adults and broods. In order to accumulate such a large amount of floral products, the bees of a given colony such a large amount of floral products, the bees of a given colony must visit not less than 500 million flowers. If we add to this the 20 to 25 kg of marketable honey and 3 to 4 kg of reserve bee bread, the total number of flowers visited and pollinated by the bees of one large colony over one season of their work can reach 600 million. The bees of an apiary consisting of 100 colonies visited and pollinated over the spring-summer period not less than 60 billion flowers of entomophilous plants. Such in reality is a miracle and biologically useful pollination work of honey bees. The count of 600 million pollinated flowers can be taken as a unit of measurement for the pollination work of a single bee colony (Kozin, 1976). Bohrat (1952) has remarked that an estimate of 80% of the insect pollination of our commercial crops is performed by honey bees. Thus, honey bees are no doubt one of the most valuable pollinators and have the advantage that they can be easily cultured and moved about at will (Romoser, 1981).

POLLINATION OF FRUIT CROPS

A brief mention may be made of the pollination of various types of fruit crops. Verma and Jindal (1997) have described a large number of fruit crops, i.e. apple, pear, stone fruits, apricot, peach, plum, cherry, almond, walnut, pecan, chestnut, hazelnut, pistachio, persimmon, olive; berrys fruits like kiwi fruits, grapes, strawberry, blueberry, raspberry, blackberry, cran berry; Tropical and Sub-Tropical fruits; mango, banana, papaya, citrus, litchi, ber, guava, sapota, jamun, loquat, fig, pomegranate etc. pollinated by several insects, in addition to self and wind pollination. In most of the cases pollination takes place through insects.

It has been pointed out in several places in the present work that insect pollination of almost all plants on this earth takes place and accordingly the flowers have also developed many special features to attract specific insects. For example, insects of various orders are attracted by colour, shape of the flower, odour or scent and food value etc. It has been reported by Gupta and Gupta (1997) that colour of the flowers furnishes visual attraction to the pollinators. In insects, vision extends from ultraviolet to yellow orange but these exhibit greater sensitivity to some specific colour range. Bumble bees, *Bombus* and honey bees *Apis* have trichromatic vision and so in the moth *Deilephila.* Bees are able to recognize only yellow, blue green, blue and ultraviolet and not the red colour. Since most of the red blossoms have a blue component, these appear to bees as blue or ultraviolet. Further, ultraviolet is an important wave band, a primary colour attractant to highly evolved pollinators such as bees and even to some flies. Some flowers reflect ultraviolet and other wave bands strongly, some reflect weakly and some do not reflect at all. These reflection patterns may also change with age of the flowers which result in a colour change in the blossom and provide clues to the visitors about the age of the flowers. It has been pointed out by Kugler (1936) that in *Aesculus hippocastanum,* orange nectar guides on white petals turned red with aging of the flower. It is of common experience that the flowers which appear, extremely attractive and vividly coloured during early stages after a few days - as they grew older, their colour changes and fades away due to several environmental factors, temperature may be the one.

In addition to colour, visits of bees and other insects are also influenced by the shape of the flowers and accordingly the flowers have modified themselves to allow easy entry of the insect inside the flower for pollination, as already mentioned elsewhere in the text.

The scent or odour is another significant factor that plays a vital role in releasing instinctive reactions and attracts specific insects for pollination.

As already stated, insects visit flowers mainly for meeting their food requirements which include pollen and nectar. In addition as suggested by Gupta and Gupta (1997) some foragers also visit the flowers for perfume, food bodies, place of protection or prey mates and oviposition sites. Perfume is collected by male euglossine bees from some orchids. These bees have special perfume containers on their hind legs and the collected perfume may be utilized after some change in the form of pheromone, attractive to females, in congregation for mating (see : Dodson, 1975; Williams and Dodson, 1972).

As have been remarked earlier, the honey bees play a very significant rather vital role in enhancing the productivity levels of different crops. The non-consumptive benefits of increasing crop yield through cross pollination by honey bees are though difficult to quantify, yet these far out weight (10 to 20 times) direct value of the social insects as producers of honey and the hive products. The value of bee pollination in crop production in USA has been estimated more than 20 billion US dollars per annum, in Canada 1200 million Canadian dollars and about 3000 million dollars in European community (see : Williams, 1991). A recent FAO report indicates that in 20 developed and developing countries of mediterranean region, these estimates are about 5200 million US dollars per year. As stated, the honey bees are the most efficient pollinators of several cultivated and wild plants because of their manifold value - because the body parts of the honey bees are specially modified to pickup a lot of pollen grains, having potential for long working hours; they can adapt themselves to different and even most adverse environmental conditions.

Honey bee's foraging behaviour is a balance between maximizing individual efficiency in food collection and the use of available food resources by the colony as a whole. Nunez (1982) and Seeley (1983) reported that between 5 and 35 percent of a colony's foragers are scout bees whose role is to find previously unexploited food source and to communicate its information to nestmates through the waggle dance i.e. by their language (Lindauer 1971, Gould, 1976). Nectar concentration determines the species that will be foraging and nectar abundance influences the size of the foraging population of the species. As nectar quantity declines, the number of foragers in the hive reduces (Frisch, 1967). Foragers that find species with higher sugar concentrations dance with more vigour and recruit new foragers which pollinate more flowers. As remarked above the food is the main source of guide for foraging behaviour of bees. The foraging speed and the rate at which bees visit flowers depends on the amount of nectar and pollen (i.e. food) present, control the foraging behaviour of insects, other factors are floral structure and the climatic conditions. It is of considerable interest to note that the time spent on a flower was directly proportional to the amount of nectar present. The number of flowers visited by honey bees depend upon the type of bloom and the density of flowers on a particular cultivar. Honey bees foraging is affected by weather conditions viz. temperature, relative humidity and solar radiation are the major factors influencing the honey bee flight. In general, foraging activity of honey bees increases in response to an increase in the temperature until it reaches 38°C (Eckert, 1955) but further increase in temperature retards foraging while nectar and pollen collection is almost suspended at about 43°C (Gary, 1976).

Furthermore, the major anthophilous (Flower loving) taxa among insects are the beetles (Coleoptera), flies (Diptera), wasps, bees and ants (Hymenoptera), thrips (Thysanoptera) butterflies and moths (Lepidoptera). All these insects visit flowers primarily to obtain nectar or pollen but even some predatory insects may pollinate the flowers that they visit. Nectar primarily consists of a solution of sugars especially glucose, fructose and sucrose. Pollen often has a high protein content as well as sugar, starch, fat and traces of vitamins and inorganic salts.

Beetle pollination (Cantharophily) may be the oldest form of insect pollination. Beetle pollinated flowers are often white or dull in colour, strong smelling and regularly bowl or dish shaped. Beetles visit flowers mostly for pollen, although nutritive tissue or easily accessible nectar may be utilized.

Fly pollination (Myophily) occurs when flies visit flowers to obtain nectar although hover flies (Syrphidae) feed chiefly on pollen rather than nectar. Fly-pollinated flowers tend to be less showy than other insect pollinated flowers but may have a strong smell, often malodorous. Flies generally utilize many different sources of food and thus their pollinating activity is irregular and unreliable.

Bees are regarded as the most important group of insect pollinators. They collect nectar and pollen for their brood as well as for their own consumption. There are over 20,000 species of bees worldwide and all are anthophilous. Plants that depend on Melittophily (bee pollination) often have bright, yellow or blue, sweet smelling flowers with nectar guide-lines (often visible only in UV light) on the petals that direct pollinators to the nectar. The main bee pollinator worldwide is the honey bee, *Apis mellifera*. The pollination services provided by this bee are extremely important for many crop plants.

Flowers pollinated by butterflies and moths are often regular, tubular and sweet smelling. Moth pollination (Phalaenophily) is typically associated with light coloured, pendant flowers that have nocturnal or crepuscular anthesis (i.e. opening of flowers), whereas butterfly pollination (Psychophily) is typified by red, yellow or blue upright flowers that have diurnal anthesis. As pointed out somewhere else, most of the members of the Lepidoptera feed from flowers using a long thin and retractile proboscis (see : Figs. 35-36 A,B).

Insect-plant interactions associated with pollination are clearly mutualistic. The plant is fertilized by appropriate pollen, while the insect obtains food and/or fragrances supplied by the plant, often specially to attract the pollinator.

Further, within flowers self pollination also brings some of the above advantages but continued selfing induces deleterious homozygosity and is rarely a dominant fertilization mechanism. In general, it is advantageous to a plant for its pollinators to be specialist visitors that faithfully pollinate the flowers of only one or a few plant species. Pollinator constancy which may initiate the isolation of small plant population is especially prevalent in the Orchidaceae - the most specious family of vascular plants (Gullan and Cranstan, 1994).

It may be concluded that Cantharophily (beetle pollination), Myophily (fly pollination), Myrmecophily (ant pollination), Melittophily (bee pollination), Phalaenophily (moth pollination), Psychophily (butterfly pollination) are the principal ways of plant pollination by insects but it is the Melittophily i.e. bee pollination which plays the dominant role in majority of the plant species. A large body of information on this fascinating topic of insect pollination and pollination ecology is available in the literature. The readers interested in the subject must consult the selected bibliography for detailed information.

6

FOOD AND FEEDING BEHAVIOUR

Insects may be broadly grouped on the basis of their food material into saprophagous or detritivores, phytophagous or herbivores, zoophagous or carnivores. Insects that fall into more than one of these categories are called omnivorous (omnivores).

1. SAPROPHAGOUS

These insects feed on rotten or decaying organic materials like leaf-litter, dung etc. Such materials were probably the basic food source for the most primitive insects which lived on the forest floor. These insects are important in the progressive - degradation of decaying organic material. Cockroaches for example, subsist on a wide variety of dead plants and animal matter. They are commonly described as scavengers. Other example of saprophagous insects are dung beetles (Fig. 37A-C) and a large number of soil inhabiting apterygotes e.g. Collembola.

2. PHYTOPHAGOUS

In this case the living plants are the source of food (nutriment) for several insects (Figs. 38, 39AB, 40). Insects of the order Orthoptera, Hemiptera, Homoptera, Lepidoptera, Thysanoptera, Isoptera etc. are basically phytophagous. In addition to these numerous insects of the order Coleoptera, Hymenoptera and Diptera are also phytophagous.

One of the most important requirement of butterflies and moths is moisture rather than food, because they dehydrate very quickly. On account of this very reason in the dry weather, butterflies live much shorter lives and they can be kept alive on water alone. The commonest food for butterflies is nectar and they have distinct preference for certain flowers which provide the best food. Moths have similar feeding habits. The diet of butterflies includes protein obtained from the pollen of the flowers they visit. Some secrete a liquid on to a pile of collected pollen dissolve it and then drink the enriched liquid.

Fig. 37. **A.** Dung beetles pushing the dung ball, **B.** A dung ball, **C.** Larva of dung beetle feeding on dung ball.

Further, some phytophagous insects may show very little feeding preference while others show some degree of preference and still others are very specific as to food choice. Some insects like *Schistocerca gregaria, Amsacta* spp., *Locusta migratoria* are polyphagous and they will eat almost any green plant in their path.

On the other hand the milkweed bud *Lygaeus militaris*, `Ak' grasshopper *Poecilocerus pictus* and some others are monophagous feeding only on the plants of a single species or genus. The phytophagous insects which lie in between these two extremes i.e. monophagous and the polyphagous are the oligophagous insects. Variation also exists with regard to what part of a plant is utilized as food. There are those insects that feed in or on roots, stems, leaves, wood parts, buds, flowers and fruits (Figs. 38, 39 AB, 40).

Fig. 38. Sawfly larvae feeding on a leaf

Every insect is normally very specific regarding the part of the plant which it eats. On the basis of feeding behaviour these are termed as leaf cutters or rollers, leaf minors, rootworms, root borers, stem borers etc. In addition to providing food (nutriment) for the phytophagous insects, the plants usually serve as a place to live, specially in the case of larval and nymphal forms. Many fungus eating insects for example termites (Isoptera) maintain fungus gardens within their nests. According to Weber (1972) these are mutualistic relationships in that both the fungi and the insects derive benefit.

A

B

Fig. 39. **A.** Hawk moth larva feeding on its food plant, **B.** A nymph of *Schistocerca* feeding on crop leaves

Fig. 40. A ladybird beetle feeding on bean leaves

3. ZOOPHAGOUS

The zoophagous insects use living animals as a source of food. They may be classified into predators and parasites (Figs. 41A,B - 45).

The predatory insects capture and eat on the spot usually other insects e.g. mantids (Figs. 41B, 42). Predators are usually larger than the prey and many prey individuals are eaten during the life of the predator. The parasites live on or within the bodies of their food source i.e. host. The parasites do not immediately kill the hosts and if the host is killed only one host is normally required during the life of the parasite and a parasite is always smaller than its host, however, an immature insect parasitizing another insect may grow and attain the same size as the host.

Fig. 41. **A.** Robberfly feeding on moth. **B.** Praying mantis feeding on a grasshopper.

The orders Odonata, Hemiptera, Heteroptera, Neuroptera, Coleoptera, Diptera and Hymenoptera include a large number of predatory insects. The predatory insects normally possess speed, agility and or stealth e.g. robber flies (Fig. 41A) and dragon flies are well known for their agility, capturing the prey in the mid air. The praying mantids (Figs. 41B, 42) and the dragon fly nymphs provide

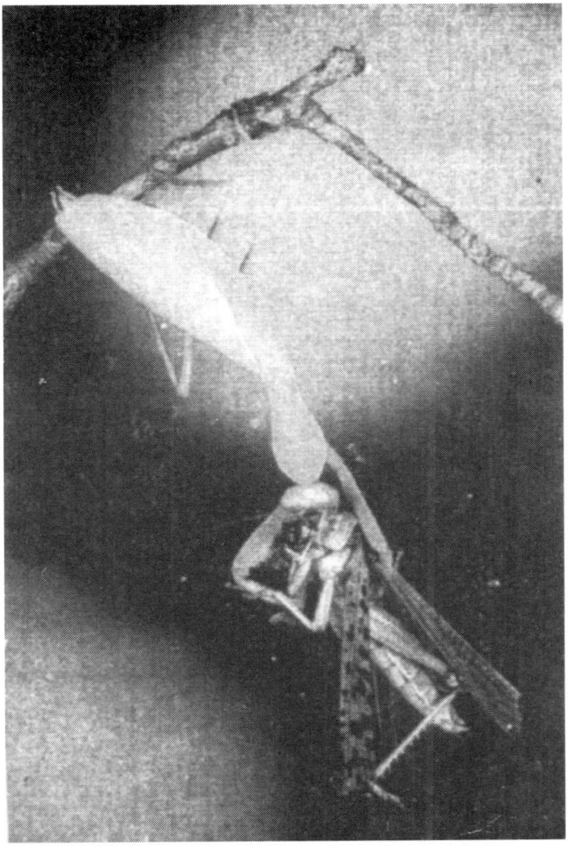

Fig. 42. A mantis grasping a locust for feeding.

examples of stealth in prey capture. These insects remain motionless and strike at prey with speed and accuracy. Some insects like antlion larvae (Neuroptera) (Fig. 43) and larvae of the caddis flies (Fig. 19 Chapter 2) (Trichoptera) trap their prey. Further, it is interesting to note that the caddis flies construct their shelter by a variety of materials (see : Fig. 19B-E chapter 2). The antlion forms a shallow cone-shaped pit in a sandy area and burries itself just beneath the surface at the bottom of the pit (see : Fig. 43). When an ant or any other small insect passes near to the pit to cause the sand to roll down the sides, the antilon responds by creating a miniature landslide. This brings the prey within reach of the antlion's jaws. The aquatic larvae of some caddis flies construct silken nets in which they are able to catch small organisms (Fig. 46).

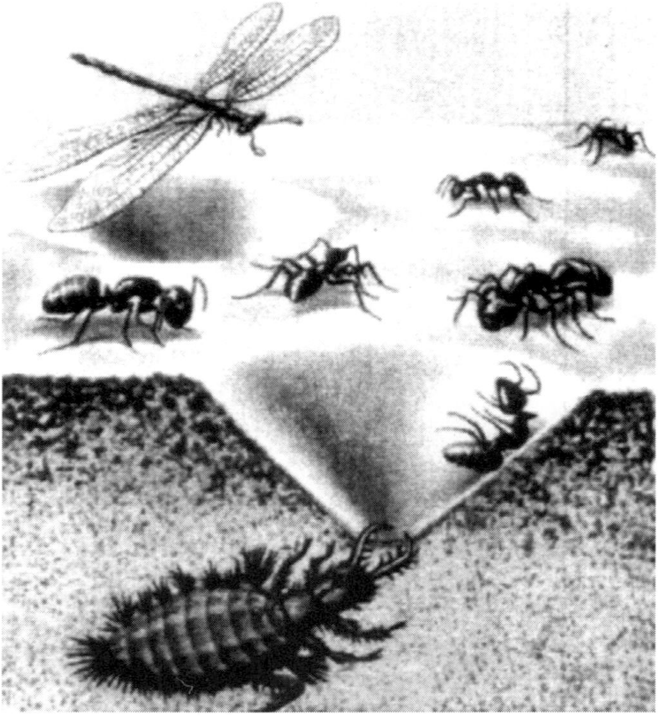

Fig. 43. Larva of antlion (in the pit) feeding on ants.

In the higher Diptera the larvae have poorly developed mouth parts and move about very slowly, thus predaceous species like many syrphid flies (Syrphidae) depend on weaker or more sluggish prey e.g. aphids, scale insects, mites, insect eggs etc. The ladybird beetles feed on aphids (Fig. 44 A,B). A large number of predaceous insects eat other insects particularly phytophagous species. The beetles of the family Carabidae, Drilidae etc. and some species of giant water bugs (Belostomatidae) feed upon snails and other insects (Fig. 45). Protozoans compose a significant proportion of the diet of many aquatic insects.

Parasitic behaviour like the predatory behaviour has arisen independently on several occasion among insects. Most insects that are parasitic on vertebrates are found in the order Anoplura (sucking lice), Mallophaga (chewing lice), Heteroptera (bugs), Diptera (flies) and Siphonaptera (fleas). The ectoparasitic insects derive all or part of their nourishment from the host and live entirely on its external surfaces. On the other hand, the endoparasitic insects invade the tissue of the host. The majority of ectoparasite insects of vertebrates are blood feeders i.e. haematophagous. The haematophagous insects like adult female mosquitoes, nymphal and adult bed bugs, as well as reduviids, visit host vertebrates only to take a blood meal and exhibit varying degrees of host specificity.

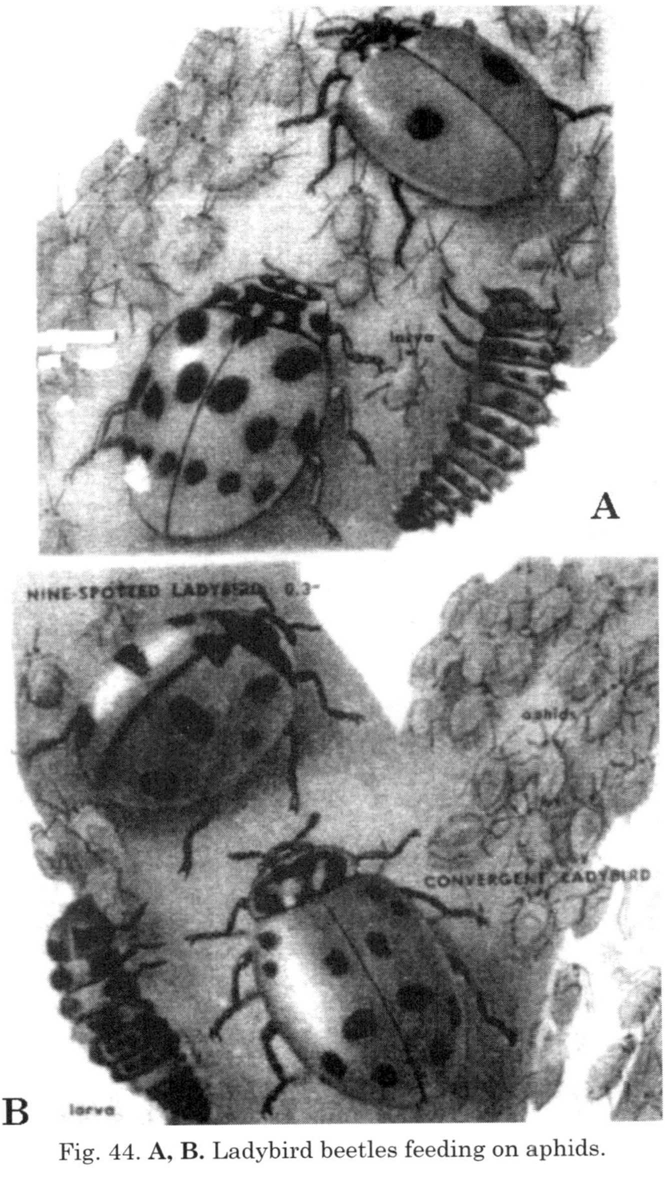

Fig. 44. **A, B.** Ladybird beetles feeding on aphids.

Fig. 45. A giant water bug feeding on a dragon fly nymph.

Fig. 46. Nest of predaceous caddisfly larva for capturing the prey

It has been found that many scorpion flies (Mecoptera) feed on the larvae and pupae of other insects (Fig. 47).

Fig. 47. Scorpion fly feeding on a caterpillar

Almost all insects that are endoparasites of vertebrates are dipteran larvae (maggots) in several families which may invade the alimentary canal, nasal cavities and open sores. These larvae attack on a wide variety of vertebrates including man. An animal infested with fly larvae is said to be suffering from `myasis'. Most insects that parasitize other insects do so as larvae and ultimately destroy or kill their host are known as parasitoids. Majority of the parasitoids live and feed as larvae within or on the host and become free living adults. Most of the parasitoids belong to order Hymenoptera and super families Icheumonoidea, Chalcidoidea, and Proctotrupoidea. Several dipteran families also contain parasitoids e.g. Tachinidae, and Sarcophagidae. These flies attack a wide variety of insects e.g. grasshoppers, caterpillars, bugs, beetle-grubs and adults. Some parasitoids exhibit a high degree of host specificity attacking a single or very few species of insects. In most of the cases only a single parasitoid can exist in a host. However, there are cases where several hundred or thousand can develop in a single host.

The parasitoids feed on non-vital tissues until they are ready to pupate and then they devour vital tissues. There is an other phenomenon the hyper parasitoidism or hyper-parasitism. In this case a parasitoid attacks on another parasitoid e.g. one may find a caterpillar harbouring a larva of a tachinid fly which in turn harbours a chalcid wasp larva. Likewise many Entomophagous (i.e. insect eating), insects, particularly the hymenopteran parasites have been used to control various insect pests (see : Awasthi, 2009 biological control).

Furthermore, it has been found that some insects particularly beetles of the family Meloidae invade the nests of solitary bees. The beetle larvae destroy the eggs of the bees and consume the food stored in the cells of the nest. This phenomenon has been called ``Cleptoparasitism".

4. OMNIVOROUS

The insects which have no food preference and eat almost any organic material are called omnivorous e.g. cockroaches. Similarly there are some insects which feed on microorganism are ``micropha-gous" and those feeding on fungus are ``mycetophagous" or ``fungi-vorous" or ``mycophagous". Thus we find that the feeding behaviour of insects largely depend on the food material which they take.

Fig. 48. Small tortoiseshell laying eggs on a food plant

LOCATION OF FOOD SOURCE

Dethier (1966) has described the various steps involved in the feeding behaviour of insects like locomotion, bringing the insect to its food, cessation of locomotion on arrival, biting or probing, sucking, continuous feeding, termination of feeding etc.

In many herbivorous insects, the female parent oviposits either on or near to a food source for her offspring (Figs. 48, 49 A,B, 50). The same is true for most of the parasitic insects. In these cases location of oviposition site, host location and food location are not distinctly separate activities. Visual as well as olfactory stimuli are probably the main factors utilized in food location by insects in general.

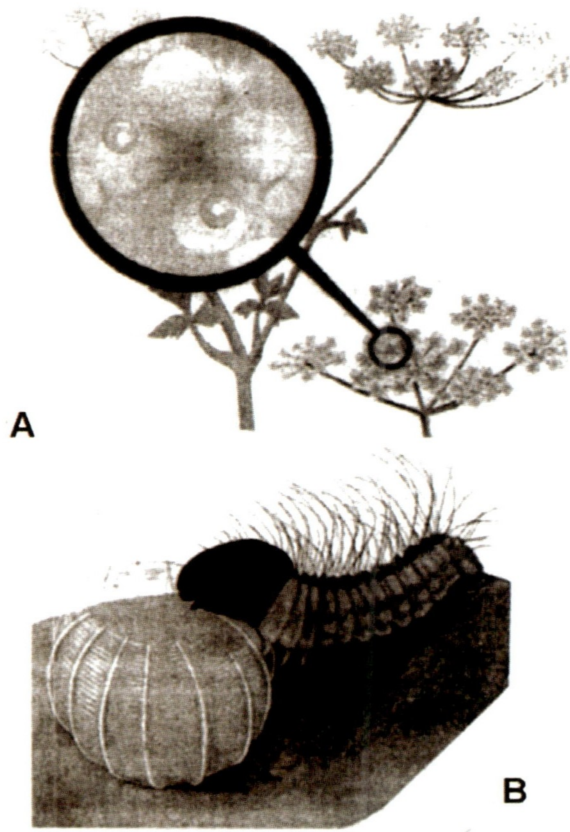

Fig. 49. **A.** Swallow tail eggs on food plant. **B.** Newly emerged larva of a butterfly feeding on egg shell.

Fig. 50. A silver silk or tailed comet moth laying eggs on the food plant of its larvae

As visual stimuli, colour form and movement in the location of flowers by honey bees are commonly used. Many butterflies also respond to these stimuli. Dragon fly-larvae and adults and praying mantids rely upon visual stimuli in prey capture. Similarly in the case of mosquitoes, tse-tse flies and other haematophagous insects, visual timuli are very important.

Volatile chemicals (as olfactory stimuli) are involved in food location in dung feeding insects such as dung beetles and many flies of the family Sarcophagidae. These insects are attracted by the volatile substances ``skatol" and ammonia, which are found in their food. These substances are referred to as *token* or *sign stimuli* since they serve as indicators of substances with nutritive value but have none themselves. The token stimuli are very important in the attraction of herbivorous insects to their host plants. Olfactory stimuli also play a role in host location by mosquitoes. Carbon dioxide, steroids, aminoacids and other volatile substances are also involved. In addition to visual and olfactory stimuli; tactile, thermal and hygrostimuli are involved in host location in ectoparasites of vertebrates.

The back swimmer, *Notonecta* (Notonectidae) locates prey by orienting itself so that it receives equal stimulation of the sensory hairs on its oar shaped legs. The larvae of ladybird beetles (Coccinellidae) catch their prey aphids by crawling along the veins of a leaf and periodically stopping and scanning the vicinity by turning their body to and fro about the temporarily secured abdomen and when they are successful in capturing an aphid they show an increase in the frequency of turns. This behaviour increases the probability of encountering additional prey and provides a good example of klinokinesis.

CONTROL OF FEEDING

After locating a food source, specific stimuli are involved in the induction of feeding. These stimuli are known as ``Phagostimulants". They may be purely token stimuli (without nutritonal value) or they may be substances that have definite nutritive value. Many of the volatile oils characteristic of certain species of plants are phagostimulants e.g. cabbage aphid (*Brevicoryne brassicae*) is induced to feed even on an abnormal host plant when the substance sinigrin, a material extracted from cruciferous (mustard) and other plants is present. Likewise mustard aphid, *Lipaphis erysimi* is also attracted. Sucrose in many insects is an effective phagostimulant. Amino acids, glucose, certain proteins, vitamin C, and several other chemicals have also been found to be feeding stimulants.

It has also been frequently observed that the larval and adult stages of some parasitic, parasitoid and predatory insect species are transported to the host or to the nest of prey species by adults of the host or prey species. Feeding is controlled by several factors other than food deprivation, like-moulting, temperature, life stage, age, reproductive stage, presence of other insects etc.

Further, I would like to add that in a recent review. Krenn (2010) has described in considerable detail the food and feeding behaviour of adult Lepidoptera along with structure, function and evolution of mouth parts using scanning electron micrographs. The readers interested in the subject must consult the above work of Krenn.

7

INTRASPECIFIC AND INTERSPECIFIC INTERACTIONS

All those organisms which include members of the same or different species, as part of each others environment or ecological communities or ecosystem comes under the intra-and interspecific interactions.

1. INTRASPECIFIC INTERACTIONS

Intraspecific relations exhibit association of similar forms or sexes, parental care, social life, cannibalism etc. The various methods by means of which the members of the same species interact are related in part to population density. The advantages appear to be associated with the relatively low as well as relatively high densities. The circumstantial evidences in which high population density may be benefited include mate finding and survival of potential predation (Andrewartha and Birch, 1954). In a typical situation where the probability of finding a mate is decreased on account of low population density the result would be fewer fertilizations and hence a lower average fertility.

Whether a given population density is benefited or not is relative and largely depends on the influence of other environmental factors. This is very clear with regard to the availability of resources such as food and shelter. The maximum population of a particular species that an area can support depends on the abundance of food and shelter facilities in that area. And when resources like food and shelter are limited, intraspecific competition may become very pronounced as the maximum population is approached and may result in the death or emigration of individuals.

It has been observed that the social association has mostly been found in army worms, aphids and dragon flies. These insects live together in large numbers for various purposes like feeding or hibernating but this overcrowding is extremely harmful and results in an early exhaustion of food material, aggregation of excretory material and high mortality by toxic effects or cannibalism. Useful intraspecific relationship brings about sex meeting and parental care viz. concealing eggs in cervices, hollow plant structures, soil, stones etc. securing them by glueing or by silken threads or by depositing them in special chambers of mud or some mimicking debris or by packing them in cocoons or ootheca. Sometimes, special nests are constructed in which mud, dung, gum tuft scales, wood pulp, fibres and wax etc. are used. The leaf rolls formed by cotton leaf roller is a common example. Many male and female beetles have been observed rolling dungballs to their nests for use of their offspring. Many wasps sting caterpillars and carry them home for their larvae. Some insects keep careful watch over their eggs and carry them on or under their body till they are hatched. Insects like mole crickets and earwigs, prepare underground chambers where they live till their eggs have hatched and take good care of their nymphs. Mostly female takes this responsibility on herself and rarely a male is concerned in this matter. The female water bug, *Sphaerodema* and *Belostoma* spp. (Belostomatidae : Heteroptera), glues her eggs on the back of males where they are lodged till hatched out. The parental care is highly manifested in social insects like bees, wasps, ants, termites etc. which have a well organized society in which division of labour and sincere devotion to duty are chief characteristics (for further details see : chapter 3).

2. INTERSPECIFIC INTERACTIONS

Interspecific interactions are of many types and they may be divided into following headings :

1. Competition

The interspecific competition takes place when the needs of two or more different species for a given resource arises e.g. food, shelter etc. coincide or when their niches overlap. In situations where two species with similar life needs are brought together, one species may be expected to have a competitive advantage and ultimately eliminate the other species i.e. no two species can occupy the same niche at the same time for very long. This is the phenomenon of competitive exclusion. The segregation of the different species in an ecosystem into different niches is explained on the basis of the operation of ``competitive exclusion". A good example of competitive exclusion is the outcome when two species of the flour beetles, *Tribolium castaneum* and *Tribolium confusum* are placed in the same container of flour (Park, 1962). If the container is maintained under conditions of high humidity and temperature, *T. castaneum* invariably wins out. However, maintaining the beetles under conditions of low humidity and temperature results in a victory for *T. confusum*. In the absence of the other, either species can be maintained indefinitely under wet and warm or dry and cool conditions.

2. Symbiosis

The term ``symbiosis" in a broad sense means a close association between two different species. The association may be mutually advantageous i.e. mutualism, disadvantageous for one of the species i.e. parasitism or advantageous for one without causing any harm to other i.e. `commensalism'.

Protozoa are common partners in mutualistic relationships with insects and those found in the gut of termites, plant louse etc. provide cellulase enzyme and thus enable their hosts to digest cellulose. Neither these protozoa nor termites can survive without the others.

There are a large number of examples of mutualistic associations between insects e.g. when ants actively care for and protect aphids which in turn excrete honey dew that is ingested by the ants (Fig. 51). Likewise, the relationships between fungus growing ants and fungi are good examples of mutualism between insects and other kinds of organisms. Additional examples include the relationships between microbes and insects relative to digestion and nutrition (see : Awasthi, 2009).

3. Parasites

Parasitism in its broadest sense includes all interspecific interactions whereby one species benefits at the expense of another. Ultimately this means that all insects that eat living plant or animal materials engaged in parasitism. The food habits of parasitic and predaceous insects may be thought of as a continuum between true parasites and predators. Truly parasitic insects are a relative minority i.e. they include lice and some fleas on birds and mammals. The most effective parasites are those that obtain energy from the host without being noticed and these parasites could, therefore, be considered commensals e.g. chances are better than 50 : 50 that parasitic mites are living in the skin pores of human beings.

Fig. 51. Formica ants feeding on honey dew and protecting the aphids

Among the bacteria that infect insects the best known is *Bacillus thuringiensis*, a spore former that has been used with some success in biological control. It is pathogenic for nearly 200 species of pest insects, particularly Lepidoptera. Viruses are obligate cellular parasites composed in part of DNA or RNA. They are classified on the basis of the part of the host cell in which they develop, the presence or absence and morphology of an inclusion body and the kind of nucleic acid (DNA or RNA) present. Some insect viruses produce granular inclusion bodies, most produce crystalline polyhedral bodies in the nucleus or cytoplasm of an infected cell. The virus particles may be alone in an inclusion body or in packets. Several viruses that are pathogenic for insects have been identified and described. Most have been found among Lepidoptera, although several also are known from the Hymenoptera, Diptera, Coleoptera and Neuroptera. It is interesting to note that a virus may act in conjuction with a bacterium to produce a disease syndrom not produced by the virus or bacterium alone or separately. Such is the case in two silkworm diseases, gattine and flacherie, *Streptococcus bombycis* and *Bacillus bombycis* respectively being the bacteria involved. Insects speceially aphids and leafhoppers are important vectors i.e. carriers of viruses pathogenic for plants. Insects also serve as vectors of several viruses that attack humans and other vertebrates notably the orthropod borne viruses including those that cause yellow fever, dengue, and the encephalitis (Romoser, 1981). Those microbes which are similar to bacteria but are without cells walls, the *mycoplasmas* have been found to cause the yellows diseases of plants like aster yellows. Several types of leaf hoppers (Homoptera) are involved in the transmission of these organisms.

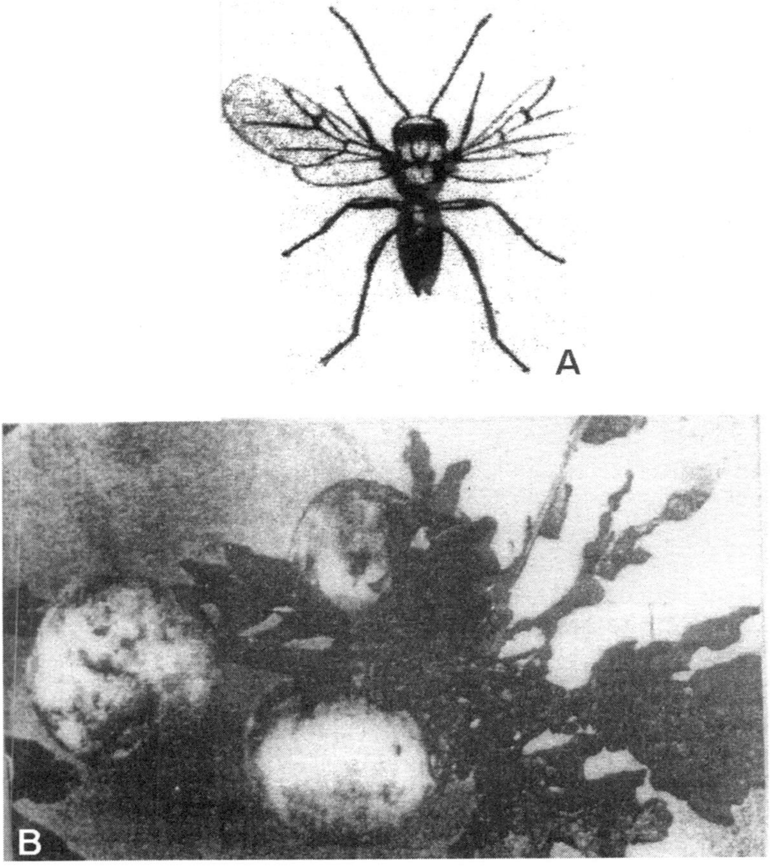

Fig. 52. **A.** Adult Oak gall wasp. **B.** Plant galls produced by insects

Some Rickettsiae are insect pathogens. In size they lie in between viruses and bacteria but unlike viruses they are capable of independent metabolism *in vitro.* The rickettsiae which are pathogenic for insects are very slow in killing their hosts. On account of slow kill of insects and potential danger for mammals they are unlikely to be used in insect control. *Rickettsia prowazekii* causes epidemic typhus and *R. p. mooseri* causes murine (mouse like) and endemic typhus which are pathogenic for humans, are vectored by body lice and fleas respectively.

Fig. 53. Marbel galls produced by wasps, *Andricus collari*

There are many species of fungi that are known to be pathogenic to insects than any other group of microbes. These *entomogenous* fungi have been described from several groups of fungi. Numerous protozoa particularly *Microsporidians* are known to kill insects but like rickettsiae and fungi, they act very slowly.

Insect pathogens have been described in the protozoan e.g. *Amoebas* (Sarcodina), *flagillates* (Mastigophora) and sporozoans (Sporozoa). The microsporodian, *Nosema bombycis* (Sporozoa) causes the ``pebrine" disease of silk worms. Other well known microsporodian insect pathogens include *Nosema apis* and *N. pyraustai,* which attack the honeybee and European corn borer respectively. In a similar manner several protozoa, pathogenic for humans and other vertebrates are vectored by insects e.g. various types of malaria (*Plasmodium* species), vectored by *Anopheles* species of mosquitoes (different forms of malaria occur in birds, monkeys, and humans). The African sleeping sickness is caused by *Trypanosoma rhodesiense* and *T. gambiae* (*gambiense*) and vectored by tsetse flies (*Glossina* spp. Diptera, Muscidae).

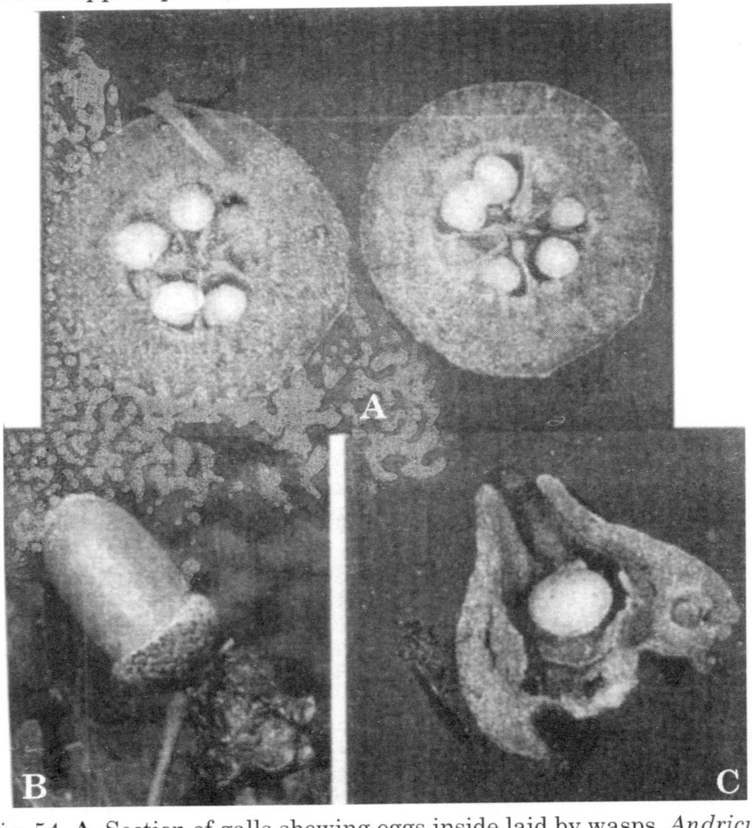

Fig. 54. **A.** Section of galls showing eggs inside laid by wasps. *Andricus collari* **B.** Knopper gall produced on acorns by *Andricus quercuscalicis* **C.** The larva of *A. quercuscalicis* live inside pearl like chamber which hatch out in spring after over wintering.

4. Predators

In the following pages a brief account on the subject is given (for details see : Awasthi, 2009). The predators are those insects which catch hold others for food. Predatory insect occur in almost all orders and most predaceous insects feed upon other insects. Such insects may be terrestrial e.g. beetles, mantids, dragonflies, antlions etc. or aquatic e.g. water-striders, water-bugs, water scorpions etc. At times, insects prepare traps cleverly for preying e.g. web, and mud pits or they may paralyse the prey by sting. The ladybird beetles (Coccinellidae : Coleoptera) both adults and grubs consume immense number of aphids. (Fig. 44 A,B). The praying mantids wait motionless and then suddenly grasp the prey in their long, forelegs (see Figs. 41 B, 42, Chapter 6). Assassin bugs are among the most expert of insect predators. About 2,500 species throughout the world display various hunting methods.

As remarked at several appropriate places, the predators of insects include insects themselves as well as other Arthropods like mites, spiders, scorpions and the vertebrates including fishes, amphibians, reptiles, birds and mammals. Before closing this part it may be mentioned that in addition to animals certain insectivorous plants such as the Venus flytrap, pitcher plant and aquatic *Utricularia vulgaris* trap, digest and assimilate insects.

5. Commensalism

``Commensalism" is a symbiotic association in which an individual benefits and the other is neither helped nor harmed. Many relationships between two or more insect species or between insects and other organisms are commensal e.g. a large number of microbes associated with insects neither harm nor help their hosts but gain food and shelter.

6. Phoresy

Phoretic relationships are good examples of commensalism. In a phoretic relationship an individual of one species attaches to an individual of another species and gains a mode of transportation e.g. some chewing lice (Mallophaga) attach themselves by their mouth parts to the louse flies (Diptera : Hippoboscidae), some of which are also parasites of birds and in this way move from one host to another.

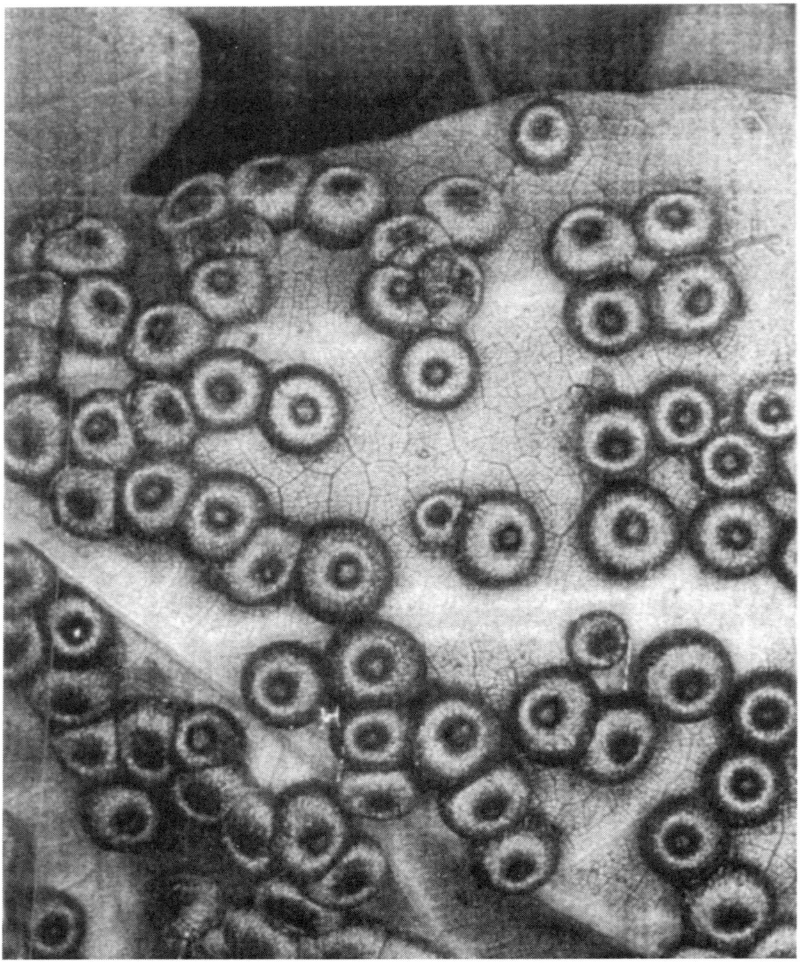

Fig. 55. Silk button galls produced by *Neuroterus numismalis* (Hymenoptera) The larvae over winter in galls and emerge in spring

Amongst the egg parasitism, hymenopterans e.g. *Trichogramma* (Trichogrammatidae) attach themselves to adult females of the host species thereby gaining immediate access to the eggs at oviposition. Thus, this phoresic relationship benefits the carried, and does not directly affect the carrier although in some cases its progreny may be disadvantaged. Phoresy provides a means of finding a new host or food source.

Some insects e.g. gall midges (Diptera : Cecidomyiidae) and gall wasps (Hymenoptera : Cynipidae) induce plants to form peculiar growths in which the larvae of these insects feed, grow and develop. These growths are called ``galls" (Figs. 52 A,B, 53, 54 A-C, 55) and are made up entirely of plant materials. Depending upon the effect of galls on a plant, such a relationship could be viewed as parasitic or commensalistic. There are several other organisms e.g. fungi, mites etc. also cause galls to develop but insects are the most common gall inducers and the most elaborate as well as specific galls are those induced to form by insects.

8

ADAPTATIONS IN INSECTS

In the case of insects, major adaptations that have evolved very early and made them most successful creatures of the universe, are the exoskeleton, small size, wings and metamorphosis. Thus, the insects in their diversity are the excellent examples of the phenomenon of adaptation. On account of the above mentioned modifications they show extensive diversification in their structure, physiology and behaviour patterns, leading to invasion of a number of habitats in space and time.

Most of the workers are of the opinion that the adaptations arise through evolution. It is interesting to note that an adaptation is a genetically based characteristic in terms of structural changes, morphological as well as behavioural patterns, and physiological approach. All these factors help an insect in survival and a successful reproductive behaviour.

The environmental factors like weather, competitors, food supply, predators etc. to which insects are normally exposed is likewise varies considerably. This variation may favour some of the important genetic combinations within populations, so that some insects will leave more offspring than others and thereby increasing the proportion of their genes in the population. The changing environment will, therefore, select genetic combinations which will be reflected in the structural, behavioural and physiological characteristics of insects i.e. their adaptations.

ADAPTATION TO ENVIRONMENT

The insects have adapted themselves to all sorts of life-terrestrial as well as aquatic and all sorts of climatic conditions extremely hot to extremely cold temperature. The success of their survival in time and space is the capability of their adaptation to any type of environmental conditions favourable or adverse. According to the mode of living place, they have modified their body structure, physiology and other needs. In accordance with the adaptational changes, they also changed their behaviour pattern and can survive at any condition of the environment, in which other animals can not survive for a few hours or so.

DESERT ADAPTATIONS

Deserts are characterised by extremes of temperature and humidity, consequently, the saturation deficiency of the air may be very great by intense solar radiation, drying winds and scanty as well as irregular rainfall. These features involve a series of difficulties for insects as well as plants too, that inhabit deserts. The problems of life are caused basically by the combination of exceesive heat and drought, the necessity to prevent internal temperatures from reaching lethal levels while restricting water loss to a minimum. The adaptations of arthropods to arid environments have been summarized by Cloudsley-Thompson (1975).

Like xerophytes, many insects have their spiracles in pits, covered with hygroscopic hairs, they are also covered with the lateral sides of the tergum, the entire body is covered with thick cuticle, black in colour, feeds mostly on stems, roots, seeds and other small organisms The insect fauna of the Indian desert is very rich and represents most of the insect orders and is well adapted to arid mode of life. The principal environmental adaptations of desert insects include : (1) size and shape, (2) orientation to the sun, (3) colour and nature of integument, (4) absorption and storage of water, (5) facultative reproduction, (6) tolerance of high temperature, (7) tolerance of dehydration, (8) specialized respiratory systems, (9) conservation of water lost in excretion and evaporation, (10) reduction in metabolic rate and (11) escape from or avoidance of unfavourable conditions etc.

Morphological Adaptations

Due to their small size and correspondingly large surface to volume ratios, very few terrestrial arthropods are able to cool their bodies by evaporation (Cloudsley-Thompson, 1964, 1991; Edney, 1977). The ability to tolerate high body temperatures, however, is an important factor in the survival of desert insects. This is reflected in the low transpiration rates from desert beetles. In arthropods, an epicuticular wax layer as well as the sclerotized chitin-protein complex of exocuticle and extension of lipid material into the underlying cuticular layers are responsible (Cloudsley-Thompson, 1993).

Dead air spaces

It has been observed that certain flightless beetles with subelytral cavities are examples of more mobile insects with large dead air spaces. Hadley (1970) has determined that the subelytral cavity of *Eleodes armata* could be upto 7°C warmer than outside air when beetles were exposed to direct sun light. He concluded that the space may function both in convective cooling and in buffering heat conducted inward. Since the configuration of the space itself is highly variable among the flightless tenebrionids, its insulative effects are probably quite different in different species.

Long Legs

Further, long legs (Fig. 56A) that raises the body of an insect well above the ground, increase its distance from the heat boundary layer next to the surface. In certain insects this combination of posture and morphology is augmented by minimal tarsal contact with substrate and therefore, minimum heat condition as well. This description is applied to ants, wasps and beetles.

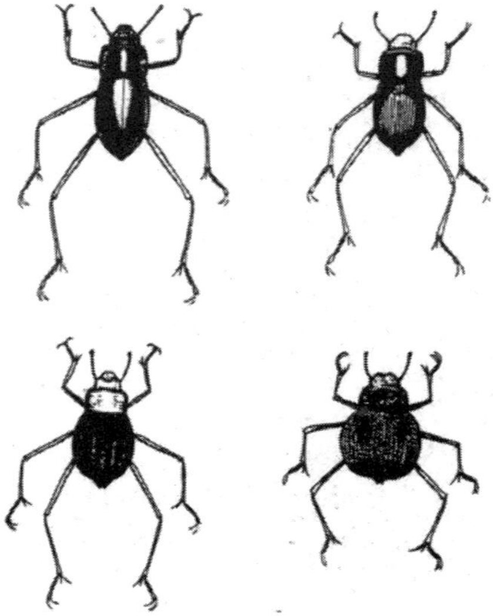

Fig. 56A. Some beetles of the family Tenebrionidae found in desert. All are rapid runners. Tarsal claws hypertrophied, terminal tibia with spines on long legs

Adaptive Colouration

Colouration is a phenomenon so important in the lives of animals including insects, that in almost every case, it has some kind of adaptive significance such as crypsis, communication, aposematism or optimizing radiation relationship. The diversity of colour so apparent amongst the fauna of less open environments is markedly absent in desert. Here, the majority of insects are either cryptic or black in colour.

Furthermore, the significance of dark colouration in desert insects holds such biological importance because so many melanistic species live in arid zones. Like colour, the integument of insects is extremely thick which prevents the evaporation of water from the general body surface. The elytra of the beetles play dual role because they also cover the whole dorsal surface of the body. They come in direct contact of sun rays so the evaporation is totally prevented from the body surface. For thermoregulation in desert insects see : chapter 9.

Some other adaptive factors are tolerance of heat and cold, avoidance of parasites and predators, hibernation and aestivation, camouflaging, secretion of repellent chemicals, migratory behaviour, reproduction at the time of rainfall etc. For details (see : Crawford, 1981). For ecological, behavioural and other adaptations of desert insects see chapter 10.

AQUATIC ADAPTATIONS

The aquatic environment can be classified into two different categories viz. the lotic (flowing) and lentic (standing). In the following pages, morphological and behavioural adaptations of aquatic insects are briefly described.

Lotic Adaptations

It has been found that those insects that live in strong water current (rheophilic species) tend to be dorsoventrally flattened, some times with laterally projecting legs. These are not strictly adaptations to strong currents being found in many aquatic insects but they do afford avoidance or minimization of exposure, by allowing the insect to remain within a boundary layer of calm water, close to the surface of the substrate.

The nests constructed by many rheophilic caddis flies assist in streamlining or otherwise modifying the effects of flow. The variety of shapes of nests, must act as broken stone against displacement.

Several aquatic larvae have suckers or convergently evolved structures that allow the insect to stick to quit smooth exposed surfaces such as rock faces on waterfalls and cascades. Silk is widely produced, allowing maintenance of position in fast flow. Black fly larvae (Simuliidae) spin a silken pad attached to a rock surface to which the posterior claws of the larva are attached. Likewise the caddisflies and chironomid midges use silk in constructing retreats, Some spin silken mesh nests to trap food brought into proximity by the stream flow.

Majority of the lotic insects are smaller than their counterparts in standing waters. Their size together with flexible body design allows them to live amongst the cracks and crevices of large stones and pebbles in the bottom (benthos) of the stream. Some live in mine in immersed wood substrates that are used by many beetles and crane-fly larvae (Diptera : Tipulidae).

Two types of behavioural strategies are commonly seen. Predatory aquatic insects frequently drift to locate aggregations of prey. Many other insects e.g. stone flies and may flies may show a diurnal periodic pattern of drift (i.e. air current or water current). Another lotic behaviour is of burrowing deep into the substrate (i.e. hyporheic zone). In the hyporheic zone, temperature and predation can be avoided although food and oxygen availability may be diminished.

Lentic Adaptations

The lentic (standing) water surface is used by many species than the lotic surface. Water striders and/or pond skaters (Hemiptera : Gerridae) are amongst the most common neustic (i.e. surface dweller) that exploit the surface film (Fig. 56 B). They use hydrofuge (water repellent) hair piles on the legs and venter to avoid breaking the film. Some staphylinid beetles use chemical means to move around the crescentic surface discharging from the anus a detergent like substance that locally release surface tension and propels the beetle forward.

Underneath the crescentic surface of standing water, larvae of many mosquitoes feed and hang suspended by their respiratory siphons as do some crane-flies (Diptera). Whirligig beetles (Fig. 57A) (Gyrinidae) also are able to straddle the interface between water and air with an upper unwettable surface and the lower wettable one. Interestingly enough, in this case each eye is divided in such a way that the upper half can observe the aerial environment and the lower half can see under water.

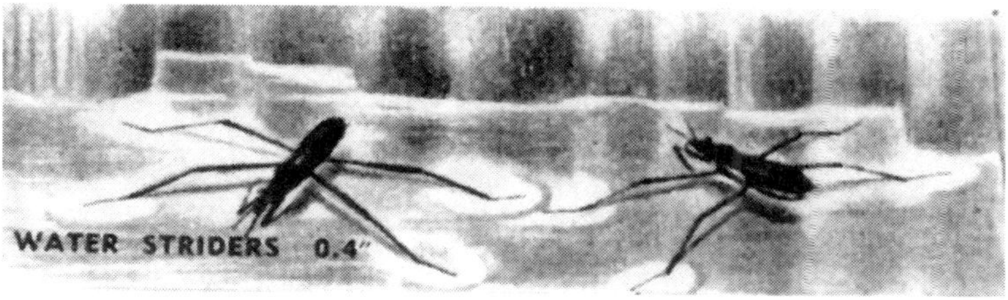

Fig. 56B. The water striders.

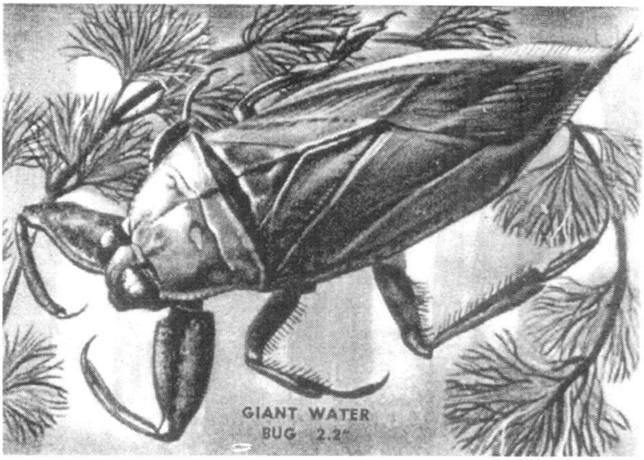

Fig. 56C. Giant water bug.

Adaptations of Body form and Gaseous exchange

Most of the free-swimming insects using their legs as a means of propulsion are adult bugs (Heteroptera) or beetles (Coleoptera). Usually their bodies are smooth and streamlined and dorsoventrally flattened so that they offer minimal resistance to movement through the water. The thrust which pushes the insect through the water is provided by the hind legs, and sometimes by the middle legs as well.

The beetle, *Acilius,* overcomes the problem by executing a very fast back stroke and a slow forward stroke but, in general, greater reliance is placed on altering the effective area of the leg. This method is employed by the giant water beetle *Dytiscus* (Fig. 57C). The hind tibia and tarsi are flattened from front to back to form a paddle, of which the area is further increased by a fringe of stiff hairs, which are articulated at the base. During the power stroke, the leg is held straight and the hairs spread to present a maximum area. On the recovery stroke the leg bends at the joint between the femur and tibia so that the paddle trails behind and the hairs fold inwards to reduce the area further. These movements are all passive and produced by the- resistance of the water: at the beginning of the power stroke, the leg and hairs are passively extended again to present their maximum area. Similarly in the whirligig beetle *Gyrinus* the segments of the hind leg are broad and flat and can be opened and closed like a fan between power and recovery strokes (Blaney, 1977).

The whirligig beetles true to their name, whirl or swim at the surfaces (Fig. 57A). They dive when disturbed and are good fliers also. Their method of carrying a bubble of air down with them at the tip of their abdomen is interesting to watch (see Fig. 59C).

The least extensive adaptations for aquatic locomotion occur in those insects which obtain oxygen from the water by means of gills or plastrons. Those insects which come to the surface to renew their air supply and submerge with a bubble of air, must swim to overcome the byoyancy their air store gives them. The gill breathers, however, have a density greater than that of water and can walk freely on the bottom. The mode of walking is essentially the same as that employed on land with some compensation being made for the resistance of the water. In the caddis fly *Trianodes* (Trichoptera) the larval case, built from plant material has a dorsal hood which produces a certain amount of lift as it moves through the water. To compensate this, the legs pull slightly downwards when walking under water. In addition to walking normally on the substrate, larval anisopteran dragonflies have developed a method of jet propulsion as an escape reaction. They normally breathe by pumping water in and out of the rectum, which contains gills. When disturbed, the insect forcibly expels water backwards thrugh the anus at a velocity of about 8 ft (2.5 m) per second and the reaction drives the larva forwards at 12-16 in (30-50 cm) per second. The legs are laid close to the body so offering a minimum of resistance to this manoeuvre (Blaney, 1977).

Water has a surface film of sufficient rigidity for insects to walk on it, provided that their cuticles are sufficiently hydrofuge, or water-repellent to prevent them getting wet. The most familiar of insects living on the water surface are the pond skaters like *Gerris* (Heteroptera). Simultaneous movements of the long middle and hind legs are used to row the insect along the surface. The tarsi appear not to penetrate the surface but rest in depressions which they produce in the surface film.

An extraordinary device for locomotion on the surface film has been developed by some staphilinid beetles of the genus *Stenus*.

They live on grass stems by mountain streams and not uncommonly fall into the water. They can walk on the surface only slowly but are often seen to glide along without apparent effort. Investigation revelas that they secrete from glands at the tip of the abdomen, a substance which lowers the surface tension of the water behind the insect. The insect is in fact drawn forward by the normal surface tension in front of it, and can control the direction of its progress by moving its abdomen from side to side.

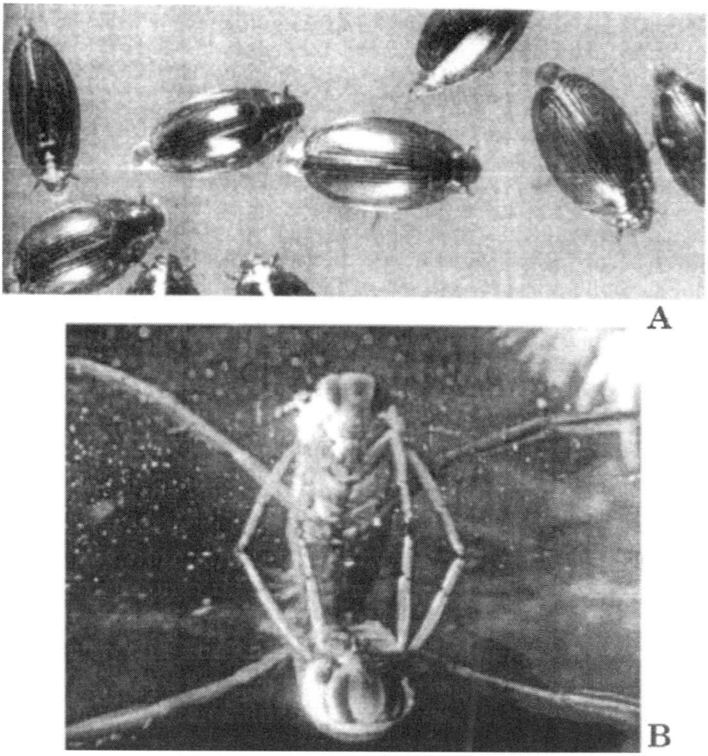

A

B

Fig. 57. **A.** Whirligig beetles. **B.** A back swimmer

Interestingly enough all the aquatic bugs are remarkable for their adaptations to life on or below the surface. Nearly all are predaceous, attacking other insects, snails, small fish, etc. In turn, these insects are food far larger fish and water birds. The water striders (Fig. 56B) taking advantage of surface tension on water, stay on the surface without breaking through. They skate along with remarkable speed. Other water bugs spend most of their time under water. Some carry down air on the surface of their bodies and use this for respiration. Other species breathe air that is dissolved in the water. The water-boatmen have an erratic swimming pattern. The backswimmers (Fig. 57B) as their name indicates, swim on their backs, but can also fly. The giant water bugs (Fig. 56C) sometimes 2 inches or more in length, prefer quiet water.

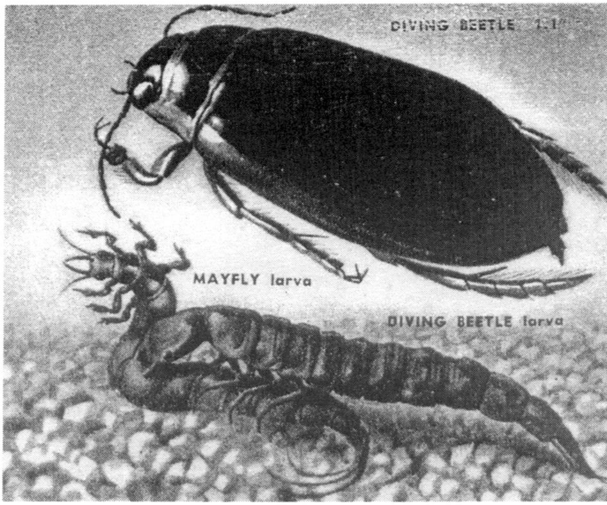

Fig. 57. **C.** Diving beetle, *Dytiscus* sp. predaceous larva feeding on may fly larva (The adult beetle carry air as a film on the underside of the body).

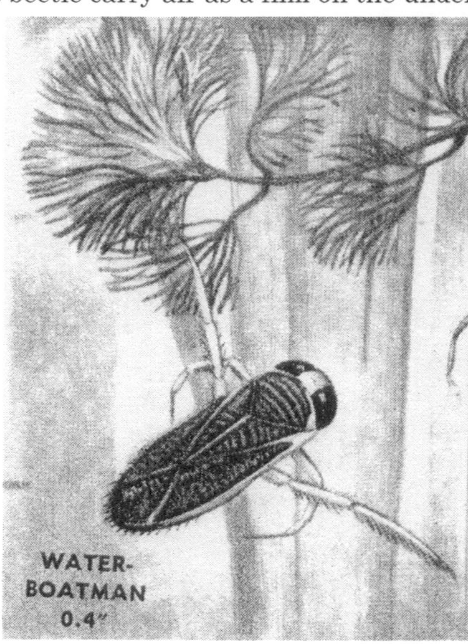

Fig. 57. **D.** Water boat man, *Corixa* sp.

Further between the water surface and bottom, planktonic organisms live in a zone divisible into an upper limnetic (where light reaches) and the deeper profundal zone. The most abundant planktonic insects belong to *Chaoborus* (Diptera) which undergo diurnal vertical migration. The diving beetles Fig. 57C (Dytiscidae) and many hemipterans like Corixidae (Fig. 57D), dive and swim actively through this zone in search of prey. The deeper zone generally lacks planktonic insects but may support an abundant benthic (i.e. bottom dwelling) community, dominated by chironomid larvae (Diptera), most of which possess haemoglobin. In the littoral zone, in which light reaches, the benthos and macrophytes can grow, insect diversity is at its maximum. Many differentiated micro habitats are available and physico-chemical parameters are less restricting than in the dark, cold and perhaps anoxic (i.e. oxygen deficient) conditions of the deeper water.

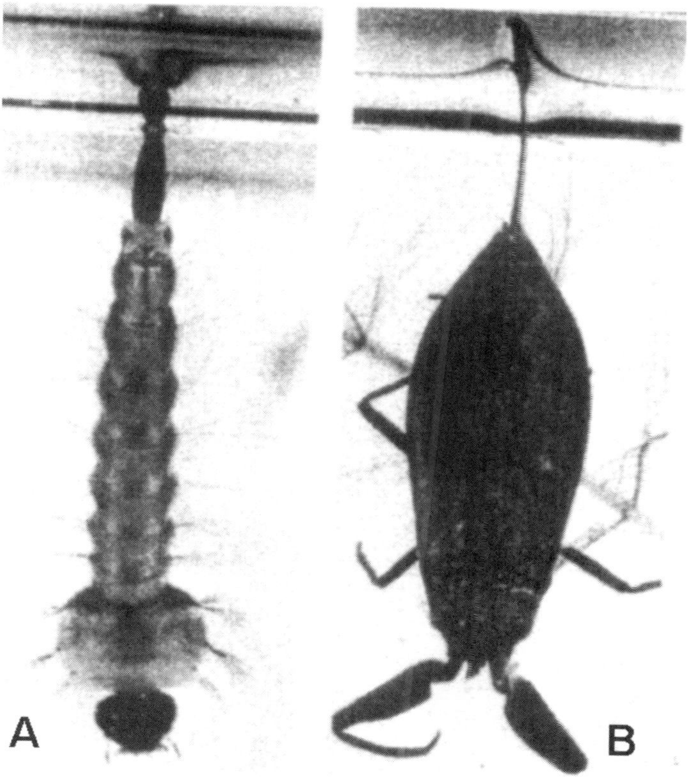

Fig. 58. **A.** Mosquito larva showing air siphon, **B.** Water scorpion with air tube

Various types of aquatic adaptations for the gaseous exchange are clearly shown in Figs. 57 A,B, 58 A, B and 59 A, B, C.

RESISTANCE TO INSECTICIDES AS AN ADAPTATION

The development of resistance to insecticides is further evidence of the evolution of the insect adaptation by natural selection. Widespread use of DDT and other synthetic organic insecticides began in 1943. In 1946, some 10 insects and mites species were resistant to chemicals that formerly killed them but by the end of 1967, this number had increased to 224, including 97 species of public health importance (Brown, 1968). In the past three decades many more insects have developed resistance to insecticides. The mechanism for developing resistance has been experimentally demonstrated in the case of houseflies, cockroaches and several other insects i.e. mosquitoes etc. Like mosquitoes that remain stationary rather than walking about on DDT treated walls,

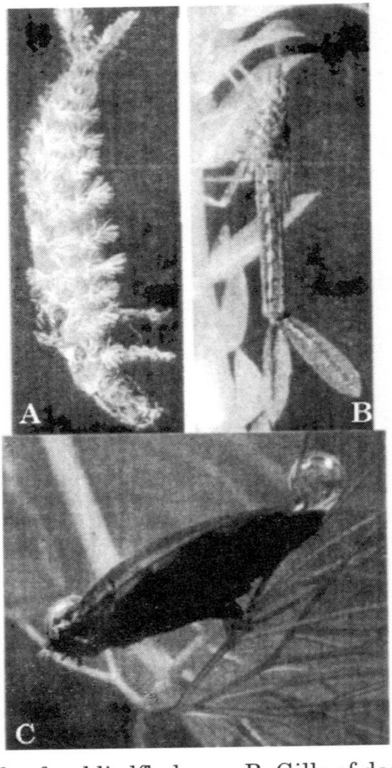

Fig. 59. A. Gills of caddisdfly larva, B. Gills of damsel fly larva, C. A whirligig beetle with air bubble in the tip of its abdomen

these resistant insects may have a genetically based behaviour that allows them to avoid contact with lethal dose and alternatively they may possess enzymes that speed detoxification of the poison. The resistant house flies can chemically change DDT to a less toxic chemical. When insecticides are applied to such a genetically variable population, majority of the susceptible insects are killed whereas the resistant remnants reproduces normally, passing on to their offspring genes responsible for resistance. The percentage of resistant insects in the succeeding generation is then much more greater. If the insecticide is applied, the proportion of resistance further increases. It has been usually taken several generations for the amount of resistance to increase beyond negligible levels, though once resistance is established in an insect population, it can often remain even after the insecticide is no longer applied.

9

THERMOREGULATION IN DESERT INSECTS

While morphological properties alone can modify thermal balance, thermoregulation usually involves behavioural and physiological modes as well. When these are used in conjunction with the fixed features of morphology, a much broader spectrum of adaptive mechanisms becomes available for an insect selecting an optimal thermal environment.

EVAPORATIVE COOLING BY BEHAVIOURAL MEANS

This form of thermoregulation is neither common nor much expected for small ectotherms in hot deserts. Species with highly impermeable integuments e.g. several beetles, do not lose water rapidly enough to bring about significant evaporative cooling, while others with high rate of cuticular or spiracular transpiration run the risk of losing body water too fast.

A case in point concerns the use of regurgitated alimentary canal fluids (regurgitated droplets) to cool the body, as found in *Sphinx* moth and honey bee. Further, the potential of regurgitated droplets for cooling bees flying at high temperatures and low humidities is considerable. The regurgitation in addition to cooling body also functions of concentrating nectar pumped anteriorly from the honey stomach. Many solitary bees and sphingid moths found in warm deserts make common use of such a mechanism. Other species inhabiting arid regions regulate evaporative cooling by controlling their respiratory transpiration.

CONSERVATION OF WATER

Although the phenomenon of evaporative cooling in insects has been observed in many species including cockroaches, grasshoppers, locusts, beetles and bees (Edney, 1977), it is not common among arthropods whose capacity to maintain water balance under stress conditions is due mainly to their ability to excrete insoluble nitrogenous waste products and the low permeability of their cuticle. In contrast, most desert insects, are without source of water and have extremely low rate of transpiration e.g. Coleoptera. Loss of water accompanying the elimination of nitrogenous waste is minimised among insects by the excretion of insoluble uric acid.

PHYSIOLOGICAL ASPECTS OF THERMOREGULATION

The metabolic rates of the desert locusts during long distance flight are comparatively low and they consequently loss less than 1% of their total thoracic heat production via evaporation. In addition, they exhibit effective control of spiracular transpiration while flying, *Sphinx* moths living in deserts appear to be less well adapted and while they possess an effective means of unloading excessive thoracic heats viz. via convective and radiative loss after its transfer to the less well insulated abdomen, 8% to 13% of this heat is lost evaporatively.

Of course, significant respiratory water loss is not a drawback if water sufficient to compensate is taken up during nectar feeding. The fact that *Sphinx* moths and bees use regurgitated moisture droplets for evaporative cooling suggests that in many cases foraging serves to rehydrate these insects to the point where evaporative cooling is not a liability. A combination, therefore, of behavioural traits and physiological controls appears to determine the thermal balance permitting flight in a number of desert insects. (see below).

OTHER ASPECTS OF THERMOREGULATION

According to Cloudsley-Thompson (1993) the thoracic muscles of insects generate a considerable amount of heat during flight but in small species the rate of convective heat loss is so rapid that little of this is retained in the muscles. On the other hand, in larger forms such as locusts, bees and moths, obligatory endothermy may raise thoracic temperatures more than 30°C above the ambient. To maintain this high rate of metabolism, the flight muscles must be provided with a continuous supply of oxygen via tracheae and tracheoles. In consequence, ventilation is accompanied by evaporative water loss which increases as the temperature of the thoracic muscles rises. The higher the temperature, the most evaporation takes place and the greater the amount of cooling. Reduction of temperature also takes place through the transport of heat by the haemolymph from the thorax to the abdomen when there is a large thermal excess. Physiological limitations place great importance on behaviour, that improves thermal and water stress. The locust's relatively low metabolic rate in flight, together with nervous control of the spiracles, appears to make this insect well adapted to the desert (Clousdley-Thompson, 1993).

ADAPTATIONS OF SOME SOIL ASSOCATED CARNIVOROUS INSECTS

Some carnivorous insects effectively use the soil just beneath the surface for protection and concealment while others wait for prey to come close enough to catch after a short run. Larvae of the neuropteran, *Myrmeleon* are an example of the first group. They and their larvae dig a concial pit in sandy soil and prey falling over the edge are quickly grabbed and over powered by these ant lions at the base (Fig. 43). Digging is accomplished in *M. immaculatus* by flicking sand as the insect circles along the bottom of a digging groove.

Another localized group of predators in desert soils consists of immature insects that depend in part on appropriate placement by the female parent in order to feed on larvae of other species. Meloid beetles provide examples of this behaviour.

Larvae of bombyliid bee-flies are also widely distributed in deserts and are often placed in the vicinity of prey by the female which flips its eggs into the burrow of a victim. Hosts of bee flies include grasshopper egg pods, as well as larvae of ant lions, beetles, moths and mining bees.

Some species of carabid beetles are also found in deserts where their predaceous larvae seem to inhabit only extremely moist soil.

DIURNAL RHYTHMS AND THERMOREGULATION

The majority of cryptozoic animals are active at night and spend the day time in their sheltered habitats. Most of them lose water rapidly by transpiration and so cannot withstand the heat and drought of day in the open. They can only afford to come out at night when the temperature falls and relative humidity of the air increases. At the same time, this is when they get rid of surplus water absorbed during the day.

The nocturnal habit confers several other advantages even to animals that resist water loss from transpiration. Enemies are more easily avoided, food more easily obtained and competition reduced. As a result of this vulnerable and primitive forms specially tend to be nocturnal. Thus among insects, silver fish, cockroaches, stick insects and crickets are typically nocturnal. More highly evolved insects are quicker-moving diurnal and live in hot, light environments. So it is probable that their nocturnalism is correlated with ecological factors such as the avoidance of vertebrate predators, rather than with physiological requirements (Cloudsley Thompson, 1967).

Of the several factors, light is undoubtedly the most important. Experiments with many species of insects have shown that with increasing the intensity of illumination, the circadian period is shortened in day- active animals and lengthened in nocturnal forms. In this way circadian rhythms can be shifted to compensate for seasonal changes in day length.

SEASONAL CHANGES

Seasonal changes have also been observed in the humidity reaction of the common earwig, *Forficula auricularia*. In summer, the insects show a clear and strong preference for the drier side of a humidity choice chamber apparatus but in winter, they respond positively to moist air which causes them to burrow into the soil for hibernation.

The duration of daylight may play a great role in controlling seasonal rhythms in insects. It may also initiate diapause, a dormant state that enables many organisms to persist in inconstant environments and regions that would otherwise be unfavourable for permanent habitation. Diapause is under hormonal control (see: Awasthi, 2009). It is usually characterized by reduced metabolism and enhanced resistance to climatic factors such as heat, cold and drought. In the case of phytophagous insects, the onset of diapause frequently coincides with some distinctive phase in the growth cycle of the host plant, yet experiments have revealed that there may be no casual connection between these events. Diapause is usually triggered by the changes in photoperiod but it is independent of light intensity provided that this exceeds a threshold value greater than that of moonlight (Cloudsley-Thompson, 1967).

10

ECOLOGICAL, BEHAVIOURAL AND OTHER ADAPTATIONS OF DESERT INSECTS

Whitford (2002) has remarked that when we think of `deserts' we think of hot and dry climates. Certainly hot and dry are good adjectives to describe deserts in a very general way. Operationally there are many aspects of `hot and dry' climates that must be considered when attempting to understand the responses of plants and animals and ecosystem processes to desert climate. Bioclimatic regimes are generally determined by temporal and spatial averaging of weather patterns. Regional climate establishes the limit for local climatic variation and determines the structure of plant communities. Local weather variables drive processes such as primary production and nutrient cycling. These variables include seasonality of rainfall with respect to seasonal temperatures, storm intensity-duration relationships, frequency of precipitation events, frequency-event size relationships, return times of particular rain events, wind-season and intensity, daily patterns of microclimates, rainfall-potential evapotranspiration relationships. Episodic events, such as extended periods of drought have long term effects on the structure and function of desert ecosystems (Whitford, 2002). In this chapter, ecological, behavioural and other adaptations of desert insects have been briefly described which are based on a comprehensive work of Whitford (2002).

MICROCLIMATE

It is important to note that most of the desert organisms avoid conditions that are stressful by selecting environments with benign microclimates. Even in extreme ambient conditions, there are locations where conditions are moderate.

The most important determinant of climatic stress in subtropical deserts in mid-summer is the intense solar radiation. The generally low humidity and lack of closed canopy vegetation cover results in little light scatter and back reflection absorption in the atmosphere. Consequently exposed surfaces receive almost all of the radiative energy that reaches the top of the atmosphere without attenuation. Such an intense radiative environment increases the importance of structural features that produce moderate microclimates.

In arid environments moving from the surface of the soil, which is exposed to direct and intense solar radiation, to 15 cm below the surface, eliminates both thermal and hydric stress. Where soil surface temperatures reach or exceed 60°C at mid-day, temperatures at 15 cm and 45 cm below the surface fluctuate only 1-2 degrees around 30°C. As the air relative humidity at the soil surface or under a litter layer drops from 40-50% shortly after dawn to less than 10% at mid-day, the relative humidity of the air in soil interstices at 15 cm to 45 cm depth is consistently between 90% and 100%. These changes in water content of the atmosphere are also reflected in the water content of leaf litter on the soil surface under a shrub canopy. Litter loses water rapidly as the air temperature increases. The intense radiation in deserts produces marked differences in the daily and seasonal maximum and minimum soil surface temperatures. At 15 cm depth, the difference between daily maximum temperatures and minimum temperatures ranges between 3°C and 7°C whereas at the soil atmosphere, the daily difference between maximum and minimum temperatures may vary between 25°C and 50°C.

Because the atmosphere over arid regions contains little moisture, solar radiation is intense and there is little `greenhouse' effect to keep long-wave re-radiation at or near the soil surface. As a result, thermal loads are high and daily thermal amplitudes are large. Temperatures of objects on the soil surface rise rapidly after sunrise and drop quickly after sunset. For many organisms the temperature at the soil surface is the most important thermal feature of the environment. Soil surface temperature quickly rise to 60°C or above in mid summer and even in mid winter may rise to nearly 30°C (Whitford, 2002).

ADAPTATIONS OF DESERT ORGANISMS

It is important to point out that the desert inhabiting organisms share most genetic and physiological traits with closely related species living in mesic environments. The suite of adaptive physiological traits of desert species are those that allow the organisms to remain within survivable thermal energy balance and water balance. It is also important to understand that many species of desert plants and animals are living close to their limits of tolerance for one or more environmental variables.

Many desert animals are able to behaviourally seek out benign microclimates. For most, but not all organisms, tissue temperatures greater than 42°C are lethal, hence it is necessary to maintain tissue temperatures <42°C. At tissue temperatures lower than the optimum for a given species, rates of reactions decreases by a factor of 2 for each drop of 10°C (the Q_{10} phenomenon). Animals that are capable of maintaining temperatures close to optimum for longer periods of time have an advantage in obtaining necessary resources. High temperatures and dry air increase water needs thus requiring behavioural and/or physiological mechanisms for reducing water loss.

There are several general categories of adaptations of both plants and animals that contribute to the success of many species of hot, arid, environments. These include: (1) avoidance of extremes, (2) physiological and morphological characteristics that reduce water losses, and thermal stress, (3) rapid response to availability of water and nutrients (4) reduced metabolic rates, and (5) mechanisms to conserve essential nutrients.

AVOIDANCE OF EXTREMES

The most successful adaptation for life in a desert or any extreme environment is for a species to avoid those conditions that exceed the limits of tolerance for life.

Most desert insects exhibit some kind of avoidance behaviour. Some like the desert cockroaches, *Arenivaga* spp. escape harsh surface conditions by burrowing 20-60 cm into the sand. At night these insects move to the surface. The timing of that activity rhythm may be triggered by the level of the temperature inversion in the soil (Cloudsley-Thomspon, 1975).

Species of insects that live in or on vegetation move up and down in the canopy throughout the day to take advantage of moderate microclimates. This behaviour has been reported in grasshoppers and probably occurs in many species of plant bugs, plant hoppers, etc. that are common inhabitants of desert perennial plants.

Various kinds of beetles are among the more common insects encountered in deserts. Tenebrionid beetles exhibit diurnal patterns of activity and seasonal activity patterns (Smith and Whitford, 1976). Some species are nocturnal and others limit activity to early morning and evening hours (Cloudsley-Thompson, 1975).

Among the most abundant arthropods in all hot deserts are the social insects, viz. termites and ants. These insects avoid extreme conditions by living in below-ground nests which may have chambers extending 2-3 m below the soil surface. Termites and ants limit their surface activity to non-stressful times of the day. Subterranean termites frequently limit surface activity to periods following rains. Species of liquid-feeding ants that continue foraging at mid-day remain above the soil surface in the vegetation when soil surface temperatures are above lethal limits. Subterranean termites use thermal shadows in the soil as cues to locate large items of potential food during periods when hot, dry conditions curtail surface activity. By feeding in large objects like dung pats and *Yucca* sp. logs, subterranean termites avoid the harsh conditions. Subterranean termites forage in litter accumulations at the soil surface in the early morning and evening when temperatures are below 30°C and the litter is moist (Whitford, 2002).

PHYSIOLOGICAL AND MORPHOLOGICAL ADAPTATIONS

As with behavioural characteristics there are suites of morphological and physiological characteristics that contribute to the `biological success' of species inhabiting deserts. Each successful species shares some adaptive characteristics with a number of other species. Successful desert inhabiting species may or may not have some unique adaptive characteristics. Physiological adaptations that are widely shared traits of groups of desert organisms have been documented here. Suites of adaptive physiological, morphological and behavioural characteristics of some desert-dwelling taxa have also been briefly described in the following pages.

Although there are few morphological or physiological specializations of invertebrates that are clearly different in desert dwelling species in comparison to related species inhabiting mesic environments, there are some that have been studied and which deserve mention. Tenebrionid or darkling beetles are numerous and conspicuous members of the invertebrate faunas of deserts throughout the world. For example, the Namib, in southwest Africa is reported to support 90 genera. Desert tenebrionids are flightless with fused elytra enclosing an air space called the subelytral cavity. Most mesic species have functional wings and moveable elytra. Studies of Cloudsley Thompson (1964) and Ahearn and Hadley (1969) demonstrated that the subelytral cavity was an important morphological adaptation that reduced water loss by evaporation from the integument. They hypothesized that evaporation was decreased because the abdominal spiracles open into the relatively humid subelytral cavity rather than to the outside air. By opening into the subelytral cavity, the spiracles are protected from the drying effects of wind. The subelytral cavity also reduces the vapour pressure gradient between the tracheae and their external environment thereby reducing evaporative water loss via the spiracles.

An alternative hypothesis was offered by Fiori (1977) and Slobodchikoff and Wisman (1981) who suggested that the subelytral cavity allows abdominal expansion which may be important for species that feed on detritus. The low nutritional value of dead plant material (detritus) requires the animals to ingest large volumes of material to meet energy needs. Thus, an expandable volume digestive system would be adaptive. An alternative advantage of an expandable abdomen is the capability to store large quantities of water in the abdominal cavity. All of these characteristics of the subelytral cavity are adaptive for detritus feeding tenebrionid beetles that inhabit hot arid environments (Draney, 1993).

It is interesting to note that the tenebrionid beetles in the Namib generate a wax bloom (wax extruded from the pores in the elytra that forms a powdery or filamentous covering of the surface). Wax bloom is produced in response to high temperature and low humidity. The wax bloom protects diurnal beetles against the high temperatures and radiant heat loads. Wax blooms may occur in tenebrionid beetles in other deserts but this has not been reported in the literature.

Perhaps, the most interesting adaptation to desert environments is cryptobiosis. Animals that enter a cryptobiotic state exhibit marked reduction in body water content and extremely low metabolic rates. When water becomes available, these animals rehydrate quickly and return to normal activity. Cloudsley Thompson (1991) reported one example of the larvae of an insect that exhibits cryptobiosis. The larvae of a chrinomid midge, *Polypedilum vanderplanki,* inhabit shallow rock pools in the semiarid regions of Nigeria. The rock pools fill and dry several times during the rainy season. When the pools fill, the desiccated larvae absorb water through their cuticle and begin feeding within an hour. Some Australian chrinomid larvae may also tolerate complete dehydration and rehydrate with no apparent physiological damage (Cloudsley - Thompson, 1991).

HERBIVOROUS INSECTS

In most arid and semiarid regions, grasshoppers make up the most abundant and largest biomass of herbivorous insects that chew foliage. There are other groups of insects that feed on foliage and on occasion these insects can be locally abundant. One example of such insects are the chrysomelid beetles (*Zygogramma tortusa*) which occasionally reach outbreak population size and defoliate a large percentage of their preferred host plant, tarbush, *Flourensia cernua* (Estell *et al.,* 1996). However, chrysomelid beetles and other leaf chewers generally make up only a small proportion of the insect herbivore biomass.

Otte and Joern (1977) reported that grasshoppers that feed primarily on annual plants, are forced, during times of absence or low densities of annual plants, to feed on perennials like *Larrea* and *Franseria* genera with high concentrations of feeding deterrents and toxins. The last plants to be affected by drought are shrubs and trees. Switching from feeding on grasses and herbaceous plants to shrubs may be necessary for grasshoppers to remain in water balance during droughts. Shrub and grass-inhabiting desert grasshoppers tend to be food specialists and to occur at low densities. Such species are generally cryptic and infrequently taken by predators. The more generalized soil loving species like *Trimerotropis pallidipennis,* also tend cryptic by matching the roughness and colour patterns of the soil (Whitford, 2002).

Shrubs may also be pruned by insects that utilize dead stems as oviposition sites. Pruning can result in compensatory growth by a shrub. Compensatory growth was demonstrated in a study of the effects of twig-girdling beetle (Family Cerambycidae) on mesquite (*Prosopis* spp.). Cerambycid beetles chew of complete girdle through the bark, cambium and phloem of terminal branches of mesquite. The girdle cuts off the supply of water and nutrients and the mobilization of sap to the portion of the stem above the girdle. The beetles oviposit in the girdled stem thereby avoiding the defense mechanisms of the plant that would encase the developing larva in a resinous tomb. The larva feed on the dead wood of the girdled stem.

CANOPY INSECT COMMUNITIES

There are many similarities in the insect faunas on desert plant canopies. The assemblages of insect populations on desert plants vary according to plant type. Perennial plants are `apparent' meaning that they are always present and a seemingly reliable source of food for herbivorous insects. Because of this `apparency' perennial plants must protect their leaves and to a lesser extent their roots and stems from being consumed by herbivores. Foliage may be protected by toxins, such as alkaloids, feeding inhibitors, such as tannins, or simply have a C : N ratio that is too large for most foliovores to survive on (i.e. quality is too low).

Despite the feeding deterrent chemistry of the foliage, *F. cernua* supports a diverse insect fauna dominated by sucking herbivores. Chewing herbivores mostly chrysomelid beetles and Lepidoptera accounted for 12-22% of the arthropod biomass. Sucking herbivores mostly Homoptera, coccids, and Thysanoptera, thrips, were 47-63% of the biomass, omnivorous ants contributed 1-20% of the biomass, and predators (wasps and spiders) varied between 2% and 14% of the arthropod biomass (Schowalter, 1996). The insect fauna of two species of snakeweed (*Gutierrezia* sp.), a small shrub that is toxic to livestock and has a high concentration of phenolics is dominated by sucking herbivores (Homoptera : Acanaloniidae, Aphidae, Cicadellidae, Eriococciddae, Membracidae, Pseudococcidae, and Psyllidae) and chewing herbivores (Coleoptera : Chrysomelidae, 12 species and Curculionidae, 12 species). (see Whitford, 2002).

An example of the diversity of the insects that utilize desert shrubs as a food source or as a hunting substrate (predatory insects) was provided by a qualitative studiy of the faunas associated with North American and South American species of *Larrea*. There were notable similarities in the faunas on *Larrea* in both areas. For example both *L. tridentata* and *L. cuneifolia* supported one or more species of cryptic grasshoppers (branch mimics and or light-dappled foliage colouration). Each species of *Larrea* supported several species of insects that prefer to feed on flowers, several kinds of beetles that feed on foliage or flowers (examples : tenebrionids, scarabs, and buprestids). Flowers of both species may be taken by fungus growing ants. Several species of ants tend homopterans on both *Larrea* species. Some differences were also recorded *L. tridentata* supported a walking stick (Phasmida), a katydid (Tettigoniidae) and several lepidopterans that were not represented in the fauna on *L. cuneifolia.*

Desert plants are no exception to these general hypotheses with respect to secondary compounds and leaf palatability. Herbivores that can avoid the chemical deterrents in the epidermal cells of leaves are those with piercing and sucking mouth parts that feed on phloem sap. These are primarily Homopteran insects but include some Hemiptera. In general the most abundant herbivorous insects on desert perennials are species in these two orders of insects. Sucking herbivores generally account for more than half of the numbers and biomass of insects on shrubs and surprisingly, on perennial grasses. The most numerous taxa in all reports are species of insects that feed on cell sap or phloem sap (sucking herbivores). Species of Hemiptera and Homoptera dominate the canopy insect faunas (Whitford, 2002).

Sucking herbivorous insects are generally the most abundant taxa on desert plants but occasionally large numbers of chewing insects dominate the fauna. The predominance of sucking herbivores feeding in the foliage of desert shrubs is usually attributed to the concentrations of feeding deterrents (terpenes) or toxins in the epidermal cells of the leaves. However, data from current studies clearly show that the foliage feeding insect fauna of grasses has the same general composition as the fauna found on shrubs (Whitford, 2002).

FEEDING CATEGORIES OF INSECTS

The presence of terpenes and other feeding deterrent compounds in the leaves of shrubs is an important factor selecting for specialized, monophagous species of insects that feed on a shrub. In a study where the insect fauna was classified by feeding niche breadth where possible, 25-28% of the insect on desert *Baccharis* spp. were polyphagous, 13-16% were oligophagous, and 2-13% were monophagous. Insects were categorized as polyphagous if they were recorded from plants of several families; oligophagous if they were recorded only from plants in the family Asteraceae, and monophagous if they were recorded only from plants in the genus *Baccharis*. Of the insects collectd from plants in the genus *Baccharis* spp. 66% fed on leaves and stems, 20% fed on external or internal parts of flowers, 11% were stem or root borers or gall formers, and 3% were leaf miners. This proportional breakdown of feeding strategies of canopy insects is probably fairly general. The comparison of insect faunas on a drough-deciduous palatable shrub (*Osteospermum sinuatum*) and an evergreen unpalatable shrub (*Pteronia pallens*) showed that the most common insect on the palatable shrub was a leaf beetle *(Cassida* spp., Chrysomelidae) which fed on new growth at any season. The most common insects on the unpalatable shrub were scale insects and plant hoppers that were tended by three species of ant. Liquid feeding (homopteran tending) ants are numerous in samples of insects collected from desert plants. The most abdunant species of ants in most desert areas are species that tend sucking insects (Whitford, 2002). These species compete for rich food resources such as concentrations of aphids on *Yucca* flowers.

Population of phytophagous insects are thought to be influenced by predictability of plant resources in space and time, the diversity of plant species within a community and spatial and temporal relationships of individual plants within patches (Cates, 1981). The more predictable plant species differ in terms of their allocation of energy to chemical defenses and in the types of products that serve as defensive chemicals. Further, it is interesting to note that even polyphagous herbivorous insects exhibit preferences. Cates (1981) found that the six polyphagous species of insects. for which data were available, all exhibited preferences for the least abundant annual and perennial herbaceous species of plant in the community. In a comparison of herbivorous insects on mesquite *Prosopis* spp., the larvae of some Lepidoptera preferred mature leaves of a species with low alkaloid contents whereas other species of larvae exhibited no correlation between alkaloid content and feeding preference. In general, polyphagous herbivores prefer mature leaves despite the known or postulated lower nutritional value, lower water content, and high concentration of digestibility reducing substances like tannins than younger leaves. This pattern was without exception for polyphagous herbivores feeding on woody perennials. However, there were exceptions to this generalization in polyphagous insects feeding on annuals (Cates, 1981). Although predictability in space and time is an important variable acting as a determinant of feeding preferences in insects, those preferences are shaped by a suite of other factors including type and concentration of feeding deterrent compounds, and water and nutrient content of leaves.

Time lags in peak populations of herbivorous insects in relation to productivity of desert plants have been documented in studies of a mirid bug (*Capsodes infuscatus*) and the geophyte (*Asphodelus ramosus*) in the Negev Desert. This insect feeds on inflorescence meristems, flowers, and fruits and in some years destroyed more than 95% of the potential fruit production. The abundance of the mirid bug was related to fruit production but with a one year time lag. Years in which the geophyte produced a large number of inflorescences, the mirid bugs exhibited high per capita reproduction. The following year, the high population of mirids destroyed most of the fruit production and had low per capita reproduction (Ayal, 1994). This kind of time lag is thought to be common among desert insect herbivores that specialize in using ephemeral resources.

It is generally observed that defoliation of desert plants is a very rare occurrence. Defoliation outbreaks typically last one season. This is true of the desert locust, *Schistocerca gregaria* in the Sahel of Africa, and probably true of most other outbreak species. One species of desert shrub, *Flourensia cernua,* experiences loss of up to 30% of its foliage to lepidopteran larvae and to adult chrysomelid beetles, *Zygogramma tortuosa* (Schowalter, 1996). The frequent defoliation of this species has been attributed to the higher N content of the foliage than is found in other desert shrubs (Estell *et al.,* 1996). Much of the vegetation in deserts produces foliage even in drought years (drought resistant woody perennials). However, because drought affects the physiology of the plants, foliage insect populations may suffer dramatic reductions during stress periods.

In the Arabian Desert of Abu Dhabi, the structure of desert insect communities was studies by light traps placed in several different plant communities (Tigar and Osborne, 1999). The Arabian Desert insect communities were dominated by moths and butterflies (Lepidoptera) which accounted for 37.4% of the total insect biomass and beetles (Coleoptera) which accounted for 42.6% of the total insect biomass. Scarab beetles (Scarabaeoidae) accounted for 42.6% of the total insect biomass. One species of dung beetle, *Scarabeus christatus,* represented 33.7% of the total insect biomass (Tigar and Osborn, 1999). Scarab beetles use ungulate dung as ovoposition sites. The abundance of one species of scarab beetle reflects the specialization of that species on the dung that is most abundant in these habitats. Because light traps sample only those mobile species that fly at night, the data on the insect communities are skewed to those species. Sucking herbivores (Homopterans) fly infrequently and hence are undersampled by this technique (Tigar and Osborne, 1999).

SURFACE ACTIVE INSECTS

There are interesting similarities in the surface active insect faunas of arid and semiarid regions. The surface active insect fauna is numerically dominated by ants in most desert areas (Crawford, 1988; Tigar and Osborne, 1997). Ants accounted for more than 75% of the surface active insects captured in the extremely arid desert of Abu Dhabi (United Arab Emirates) (Tigar and Osborne, 1997). Even in areas where vegetation is sparse or virtually absent there are likely to be a few colonies of ants. In addition to ants, the insect fauna is very diverse. Ground dwelling beetles are generally the next most abundant group of insects in desert regions. The most abundant species of ground dwelling beetles are detritivores in the family Tenebrionidae. In the northern Chihuahuan Desert, the species richness and abundance of tenebrionid beetles was higher in sand dune habitats than in areas with sandy loam soils and vegetation cover of creosotebush *(L. tridentata),* mesquite (*P. glandulosa*) or mixed grasses (Crawford, 1988). The higher diversity and abundance of tenebrionids and several other taxa in the sand dune system was attributed to soil texture and the microenvironments characteristic of sand dunes. Crawford (1988) concluded that sand is a poor conductor of heat, which means that even shallow depths provide survivable temperatures on hot or cold days. In addition, the high infiltration rate associated with deep sands allows water to accumulate below the layer of intense evaporation. The sand burrowers included camel crickets, and tenebrionid beetles (Crawford, 1988). Other studies have indicated that sandy soils are important for sand-swimming cockroaches (Edney *et al.,* 1974) and tenebrionid beetles (Seely and Mitchell, 1987). Two other factors affect the abundance of detritivores in sand dune environments : trapping and burial of wind blown detritus and the availability of thermal gradients that allow detritivores to select microsites where temperatures are optimum for the microbial gut symbionts to break down cellulose (Crawford, 1988). In shrub, shrub-grass mosaic, and desert grassland habitats, tenebrionid beetles feed on the refuse piles around the nests of seed-harvesting ants or on fine litter trapped behind litter dams. In these habitats, there is sufficient vegetation to provide moderate microclimates for the beetles during the hottest parts of the day.

The taxonomic composition and relative abundance of surface-active insect fauna in desert grassland of New Mexico was very similar to that reported by Crawford (1988) for sand dune habitats in the northern Chihuahuan Desert. The taxonomic composition and relative abundance of the surface active insect faun of Abu Dhabi was also similar to that reported for the Chihuahuan Desert (Tigar and Osborne, 1997). Data from pit-fall trap sampling of surface active insects in arid western Australia showed that, excluding ants, beetles (Coleoptera) were the second most abundant insects. Orthoptera (crickets) and Dictyoptera (cockroaches) were the next most abundant detritivores.

Arid ecosystems may support large populations of collembolans. Epigeic collembolans feed on grasses and herbaceous plants and on plant litter. Temporal and spatial patterns of epigeic collembolans studies by pit-fall trapping on a watershed in the northern Chihuahuan Desert revealed no significant correlations between abundance and species diversity and either weekly or long term precipitation patterns. During the cold dry season (November through March) *Brachystomella arida* (Hypogasturidae) and *Cryptopygus ambus* (Isotomidae) were the most abundant collembolans on the watershed. During the warm wet season (July through September) and the warm dry season (May-June), the epigeic collembolan assemblages were dominated by *Seira bipunctata* and *Entomobrya* sp. (Entomobryidae). The sminthurid, *Sminthurides pumilis* was most abundant collembolan on the ephermeral lake bottom and in the mesquite shrubland at the lake periphery. The hypogastrurid, *B. arida* was most abundant in the black-grama grassland at the base of the desert mountain. The entomobryid, *S. bipunctata,* was the most abundnat species on the creosotebush *(L. tridentata*) bajada and *Entomobrya* sp. was the most abundant species on the lower slopes of the watershed. The abundance of collembolans on this watershed suggests that the importance of these primitive insects may be missed in most studies of consumers in deserts (see Whitford, 2002).

Desert ant communities share several features : numerical dominance by small generalists (omnviores), moderate densities of one or more species of seed harvesting ants, and varying numbers of species that are primarily liquid feeders or fungus feeders. In the Chihuahuan Desert of central Mexico, there were 14 species of omnivorous ants, nine species of granivorous ants, four species of predators, three species of honey dew feeders, and two species of fungus growers. Species richness varied between 11 and 26 species of shrubland and grassland vegetation types but neither abundance nor biomass were related to vegetation type. The total species richness at the Mapimi site in Mexico (32 species) was considerably lower than reported for a watershed (50 species) in the northern Chihuahuan Desert, New Mexico.

In most deserts, the seasonal activity of ant species varies with the phenological pattern of host plants of honey-dew producing insects, flowering and seed production patterns, and the availability of moderate thermal environments for predatory species and omnivores. Peak diversity and abundance of ants foraging on plants occurs when the plants are flowering. Flowers attract insect prey and provide nectar for ants. In the hyperarid Namib, there was a close relationship between mean annual rainfall (15 mm yr^{-1} to 95 mm yr^{-1}) and ant species richness. In the Namib, most of the ants are seed feeders or honeydew-nectar feeders. In the hyperarid Atacama Desert, ants are limited to ephermeral washes and are associated with the shrubs that occur in this habitat. In the North American Chihuhuan Desert abundance of active nests of seed harvesting ants was correlated with seed drop in ephemeral plants. Abundance and species richness of active nests of omnivores and liquid feeders were correlated with rainfall with several days lag following the event (see for review Whitford, 2002).

11

THERMOREGULATION BY BEHAVIOURAL MEANS

The capacity of insects to control body temperature by behavioural means has long been recognized. Fraenkel (1929) for the first time documented the dramatic behavioural thermoregulation of the desert locust, *Schistocerca gregaria*. Since that time studies on both immature and adult insects have indicated that a wide variety of insect species from several taxa routinely control body temperature by alterations of their behaviour.

The regulation of body temperature in insects depends on size, shape, body posture, body colouration, wing position, habitat selection etc. i.e. behavioural means play a great role in this process. For a comprehensive review on this subject see Casey (1981), on which this brief account is based.

Thermoregulation is distinguished from temperature selection behaviour in which an animal chooses a particular ambient temperature and body temperature follows passively. While temperature selection is clearly adaptive and virtually all animals exhibit some capacity to seek out appropriate thermal regimes, this behaviour is characteristic of poikilothermic animals. It has been pointed out that the distinction between physiological and behavioural thermoregulation in insects is often difficult because many insects can be ectothermic or endothermic depending upon the circumstances.

REGULATION OF BODY TEMPERATURE

Insects from several taxa control body temperature either exclusively by behavioural means or in combination with endogenous heat production. The level at which the body temperature is controlled may differ among different species. The degree to which insects regulate their body temperatures in the field is often described by relating the body temperature to air temperature.

Due to their economic importance, locusts have been most extensively studied (see Uvarov, 1977). In general, locusts maintain similar degrees of regulation in warm and cool habitats. Their range of thoracic temperature is high, and the level is similar to that of other large flying insects that rely on the behaviour to control body temperature. Although locusts produce fairly large quantities of heat as a by-product of flight activity, they have little capacity to regulate thoracic temperature physiologically during flight and rely exclusively on behavioural thermoregulation (Uvarov, 1977).

Odonata

Studies on dragonflies suggested that they probably regulated body temperature and Corbet (1963) divided them into two categories (i) perchers and (ii) fliers, based on presumed mechanisms of thermoregulation. The effectivenss of behavioural thermoregulation of different percher dragonflies is directly related to body size (May, 1976).

Lepidoptera

Field observations on butterflies have suggested that they probably thermoregulate behaviourally. Thoracic temperature of butterflies during flight range from a few degrees above the air temperature to as much as 15°C above ambient temperature and regulated thoracic temperature of different species is related to wing loading. In most cases, thoracic temperature is elevated behaviourally. But in several species behavioural thermoregulation occurs simultaneously with shivering (see Casey, 1981).

Diptera

Syrphid flies show a remarkable degree of thermoregulation due to their small size and low thermal inertia by a combination of behavioural movements, related to the location and frequency of perching in sun or shade, and by shivering. The range of body temperatures maintained is much lower than that of other flying insects as a results of physical factors associated with their small size.

Homoptera

Different species of cicadas also show different capacities for control of body temperature. The desert cicada, *Diceroprocta apache* and the cactus dodger, *Cacama valvata,* show a high degree of regulation while the body temperature of the periodic 17 year cicada, *Magicicada cassina* generally parallels the air temperature.

Holometabolous Larvae

Larvae of holometabolous insects do not produce heat in sufficient quantity to elevate their body temperature. Regulation of temperature by caterpillars is dependent entirely on external heat sources. The desert caterpillar, *Hyles lineata* regulates its body temperature, while another species, *Manduca sexta* from the same family, in the same habitat, does not. The larvae of *Colias* spp., while having distinct preferred body temperatures, have only limited capacity to regulate body temperature behaviourally.

Coleoptera

Tenebrionid beetles inhabiting the Namib Desert regulate body temperature entirely via behaviour. Hamilton (1975) has shown that several species of diurnal desert tenebrionid beetles confined to substrate show mean body temperatures above 40°C for much of the day.

BODY SIZE AND SHAPE

The body size is extremely important in determining the equilibrium body temperature and the rate at which it is achieved. As body mass decreases, the ratio of body surface to body volume increases and since heat exchange occurs at the body surface, a higher relative surface area will facilitate a more rapid rate of heat exchange. Other things being equal, a small insect will reach equilibrium body temperature more rapidly than a large one under similar conditions. The large insect will have a higher equilibrium body temperature than a the small one because of higher heat capacity and will be less subject to transient changes in body temperature as a result of higher thermal inertia.

Body shape also appears to be important in determining equilibrium body temperature of insects in sunshine presumably because of alterations in surface/volume ratios within a given size range. Elongate insects, such as locusts, tend to have a lower temperature excess for a given mass than compact, spherically shaped insects e.g. bees, flies etc. The evolution of a dipteran, hymenopteran, or coleopteran body shape may be associated with maximizing the temperature excess for a given mass by reducing the surface volume ratio.

ORIENTATION AND POSTURAL ADJUSTMENTS

It has been found that insects routinely vary their body orientation in response to changes in the thermal environment. During periods when body temperatures or air temperatures are below the range normally associated with activity, insects orient the long axis of the body perpendicular to solar radiation. This behaviour often called positive orientation or basking, maximizes radiative heat gain by increasing the surface area of the body exposed to sunlight. This pattern is common in cicadas, dragonflies, some butterflies, caterpillars, tenebrionid beetles etc.

In most cases the dorsal surface of the body is exposed. Locusts, however, often exhibit an elaborate basking posture (see : Fraenkel, 1929) in which the metathoracic leg and wing on the side of the body facing the sun are moved out of the way, thus facilitating heating. In addition, the lateral side of the body rather than the dorsal surface is exposed to solar radiation because the lateral portion has a much larger surface area.

Dragonflies are particularly adept at orientation with respect to solar radiation. Figure 59D illustrate several types of positive and negative postures of dragonflies (May, 1976). May's study demonstrates how these postures vary at different times of the day in relation to environmental heat load. Dragonflies tethered in the sun, unable to alter their posture, were also unable to regulate body temperature. The obelisk posture of *P. longipennis* is common among various perching dragonflies. This posture only occurs at high ambient temperatures. When exposed to a radiant heat source in the laboratory *P. longipennis* assumed the obelisk posture, causing body temperature to fall 2.5°C (May, 1976).

Studies of several workers have clearly indicated that the size and shape are important in determining the efficacy of orientation for body temperature control. Experiments by Edney (1971) indicate that a change in orientation of 90° for the tenebrionid beetle *Onymachris rugatipennis* results in a change of about 4°C. Elongate insects expose a smaller portion of the total body surface for heating, while the long axis of the body is oriented parallel to sunlight e.g. a shift from positive to negative orientation in a final instar caterpillar, *H. lineata*, results in about a 10 fold reduction in the surface area exposed to solar radiation. (Casey, 1976).

Studies conducted on behaviour of insects in field suggests that orientation in the environment is varied with respect to wind velocity as well as solar radiation to maintain body temperature, e.g. in the sphinx caterpillar, *H. lineata*, at mid day when ambient temperature is 36°C, virtually the entire population is in vertical orientation on vegetation, resulting in a mean body temperature only 2°C above ambient temperature. However, at mid day on a cool day (ambient temperature = 28°C), most of the population remains horizontal with the long axis of the body oriented into the wind. Under these circumstances the mean body temperature is 7°C above ambient temperature (Casey, 1976). Locusts also orient with respect to the wind, as well as with respect to solar radiation.

INSECT OF THE GROUND

The importance of substrate conditions in determining the body temperature of insects has been demonstrated in many studies. During mid-morning and late afternoon, ground temperatures exceed air temperatures but are below the temperatures that limit activities of insects.

Insects, such as flightless tenebrionid beetles, which are tied to the substrate (particularly those of small body size) are at a distinct disadvantage in regulating their body temperature, because they lie within the boundary layer of the surface where air temperatures are near substrate temperatures and wind velocity is low, relative to wind velocity a few centimeters above the ground. In addition, infrared radiation and reflected solar radiation from the ground are more important to the heat budget of ground dwellers. Although they warm up more quickly than insect in vegetation, by mid-morning temperatures at the ground surface become too warm and the insects must seek cooler microhabitats under ground. Nevertheless insects on the ground often exhibit a surprising degree of control over their body temperature, e.g. the stilting posture of locusts. The effectiveness of such `behaviour is related to the distance between the ground and the ventral surface of the body and therefore to the length of the legs (see Casey, 1981). It is significant that a variety of surface-dwelling desert arthropods have evolved long legs. Stilting is characteristic of the several tenebrionid beetles. Stilting is highly effective for several reasons. By stilting, the insect (1) reduces contact between the body and the ground, thereby minimizing conduction (2) moves the body into cooler temperatures as there is a sharp gradient in ambient temperature within a few millimeters of the ground and (3) enhances convective heat transfer by exposing a greater portion of body surface to higher wind velocities.

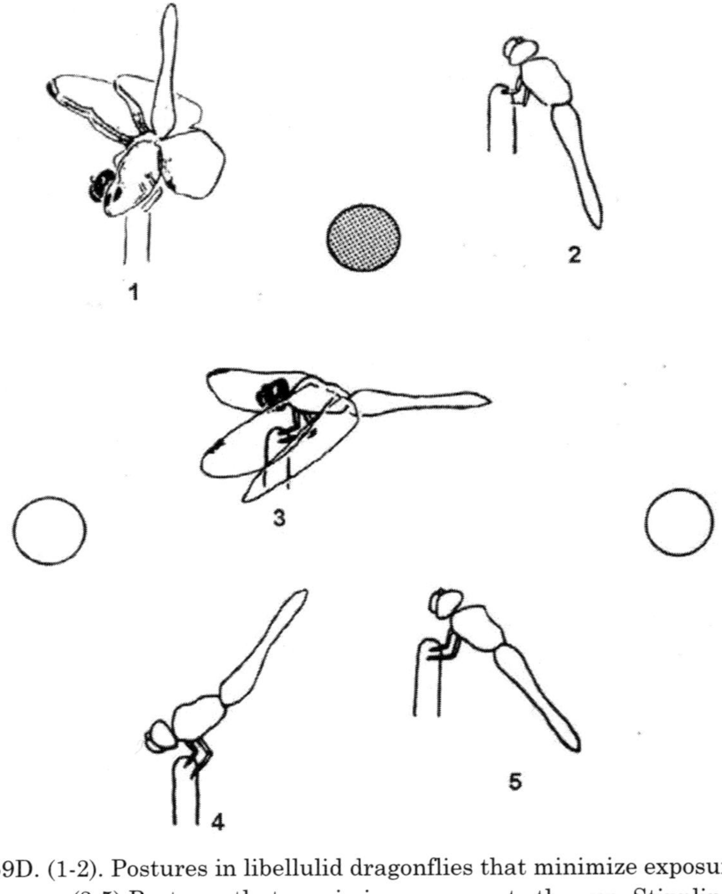

Fig. 59D. (1-2). Postures in libellulid dragonflies that minimize exposure to the sun. (3-5) Pastures that maximize exposure to the sun. Stippling indicates sun above or behind plane of paper. Typical wing positions shown in (1) and (3). (From Casey, 1981)

ROLE OF WINGS IN THERMOREGULATION

It is interesting to note that the butterflies are among the most conspicuous of the basking insects. It has long been suggested that butterflies utilize their wings to control body temperature, Clench (1966) summarized earlier field observations on the behaviour of butterlies, supplies further evidence from his own observations, and classified butterflies according to the position of their wings during basking. Dorsal baskers, as illustratyed by *Papilio polyxenes* (Fig. 59E) position the dorsal surface of the thorax perpendicular to the rays of the sun, with the fore and hindwing to the sides of the body. Under high solar loads and high ambinet temperatures the wings are folded over the body. Lateral baskers such as *Colias eurytheme* (Fig. 59F) fold the wings tightly over the back and orient the body broadside. Lateral baskers vary the leaning from side to side and roll but have little control of head up or down during basking (Watt, 1968). A third type, intermediate between the dorsal and lateral baskers is exhibited by skipper butterflies. Skipper's hindwings are spread flat to the side of the body while the forewings are perpendicular to the hind wings (Fig. 59 G).

Several studies have clarified the role of the wings in basking butterflies. The dorsal basking butterfly, *Papilio machaon,* exhibited a 30% reduction in the equilibrium temperature attained by the thorax, and a decreased rate of thoracic heating, when the wings were shaded (Fig. 59 H).

The position of the wings of basking dragonflies is important. Dragonflies of several species bask with the wings bent forward and downward and this facilitates heating, particularly over a warm substrate. In the black dragonfly, *Haganius brevistylus,* this posture is associated with rapid warming rates. The wings apprently shield the thorax, thereby reducing convective heat loss. (Tracy *et al.,* 1979).

THE ROLE OF SURFACE COLOURS

The surface colouration of insects may have important consequences for their thermal balance. Although most living tissue is essentially a black body in the infrared region of the spectrum, approximately 50% of the energy in solar radiation lies within the visible region. Consequently, different surface colours could potentially result in large differences in the quantity of radiant energy absorbed. The results of Digby (1955) have been used to argue that colouration is an important determinant of body temperature and also to discount the significance of differences in the colouration of insects.

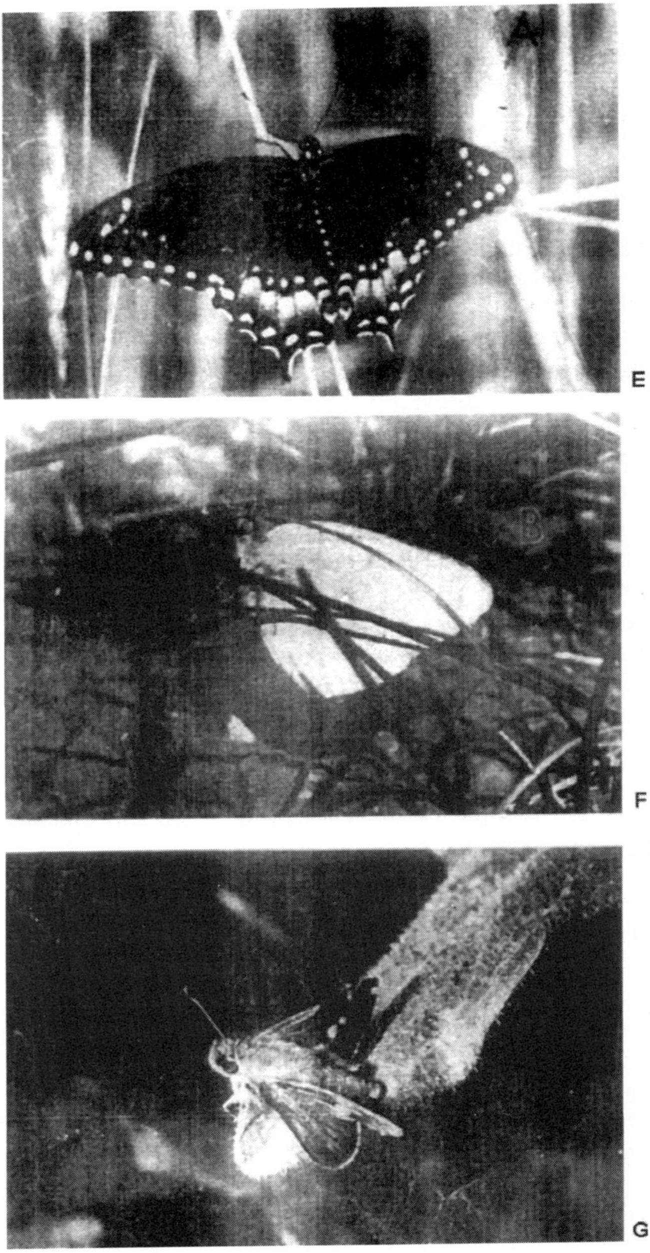

Fig. 59 (E, F, G). Basking pastures of butterflies (59E) Dorsal basking by the swallowtail butterfly *Papilio*. (59F) Lateral basking by *C. eurytheme*. (59G) Body basking by the skipper, *Hylephila phylaeus* (From Casey, 1981).

The occurrence of tenebrionid beetles having either white or jet black elytra (see Edney, 1971) in the hot, arid Namib Desert of southwest Africa has led to several studies on the thermal significance of different coloured elytra. The species having white elytra exhibit significantly lower measured abdominal temperatures than those having black elytra (Hadley, 1970). Direct measurements of reflectance indicate that white elytra absorb significantly less radiation than black ones. Hamilton (1973) has suggested that these differences in colouration relate to differences in the thermal ecology of the species, with black beetles becomes active earlier in the day because of increased radiant heat absorption, and white beetles remaining active during the hotter parts of the day because of reduced heat absorption.

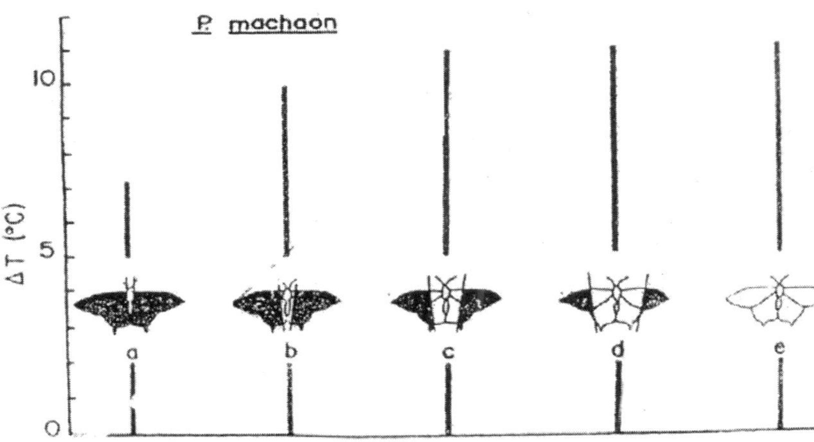

Fig. 59H. The thoracic temperature excess achieved by the butterfly *P. machaon* during basking with varying degrees (a-e) of wing shading. The thoracic temperature excess ($\Delta T = T_{th} - T_a$) is indicated by vertical lines. (Th - Thoracic temperature, Ta- ambient temperature) (From Casey, 1981).

The colouration patterns are structural and therefore, cannot be manipulated physiologically in response to environmental changes. However, their effects on body temperature are varied with appropriate behaviour. A butterfly, for example, exposes its wing bases to direct solar radiation only during basking. Under conditions of heat stress, the wing bases are not exposed to solar radiation and therefore, make little or no contribution to thoracic heating. At high ambient temperatures the tenebrionid beetle *Stenocara phalanguim* orients the long axis of its body parallel to solar radiation with its white abdomen facing the sun (Henwood, 1975). Basking cicadas expose a black dorsum, but at high ambient temperature expose a white venter (Heath and Wilkin, 1970). Consequently, structural colouration can be manipulated for regulating temperature under different environmental circumstances.

There are several reported cases of physiological colour changes in relation to temperature. Australian grasshoppers (*Kosciuscola tristis*) change colour in respose to different temperatures. Below 15°C while in a basking posture the hoppers are black, while at higher temperatures the surface colour becomes blue. The response is related to the migration of pigment granules in cuticular cells in response to temperature.

SELECTION OF MICROHABITAT

Perhaps the most extensively studied case of habitat selection for thermoregulation is that of migratory locusts (see: Uvarov, 1977). In the early morning before dawn locusts are perched on vegetation with their backs facing east. Air temperatures at this time are usually sufficiently low hence the locusts are completely inactive. Their position above the ground allows them to intercept the rays of the sun several minutes before it strikes the ground. Air temperatures rise more rapidly than ground temperatures in the early morning and higher body temperatures are achieved by basking on the vegetation. After the ground has warmed to higher temperatures about midmorning, the hoppers move from the vegetaton to the ground.

When the body temperature is elevated, flight or other activity commences. During midday, if ambient temperature is high, activity is reduced and animals take up positions high in the vegetation where increased convection and reduced solar radiation prevent overheating. Locusts also seek shade at very high ambient temperatures and remain on the ground throughout the day on cool or cloudy days.

In a few cases, the presence of insects themselves alters the microhabitat sufficiently to affect their body temperature. For example locusts (both gregarious and nongregarious species) often form quiescent aggregations comprising from a few to several hundred individuals. The aggregations are so tightly packed that individuals with a group are often in direct contact with several other locusts. This behaviour has obvious thermoregulatory significance, because these groups are invariably located in an open habitat on bare ground which is significantly warmer than the air temperature. After a locust becomes settled in the group, it usually assumes a basking posture. If the ground group is partially shaded, locusts located in the shade will move to a new, sunny location, while the animals in sunlit position remain equiescent. Group basking should result in higher equilibrium body temperatures than those obtained by individuals, because the surface/volume ratio of the ground is much smaller, than the sum of the surface/volume ratios of its individuals and convective heat exchange is reduced.

Some arctic insects bask in the flowers of *Dryas intergrifolia* and *Papaver radicatum*. These flowers track the sun throughout the 24 hr of daylight at Lake Hazen, Ellesmere Island (81°49' N latitude), acting as parabolic reflectors in focusing heat from the sun onto their sporophylls (Hocking and Sharplin, 1965; Kevan, 1975). Mosquitoes and flies perched on the flowers attained body temperatures of 6-15°C above ambient temperature and spent up to 13 min in each flower while foraging for nectar. In flowers without petals, the temperature excess of the insect is lower than in normal or desporophyllate flowers (Kevan, 1975).

In most cases of microhabitat selection, insects maintain a relatively constant body temperature by positioning themselves in thermal conditions where they are above the air temperature. If ambiant temperature exceeds the normal activity range, insects of many species seek shade and activity ceases until the air temperature declines. Once in the shade of course body temperature will follow air temperature unless evaporative cooling occurs. Consequently, particularly in very hot environments, insects having access to shade may well be poikilothermic during midday. At high ambiant temperature several desert arthropods burrow in the sand (Hamilton, 1973; Henwood, 1975) although it is unclear to what extent activity and thermoregulation occur in surface dwellers that burrow at midday to avoid heat stress.

However, locusts are also able to orient perpendicular to a radiant heat source in total darkness (Volkonsky, 1939) hence both visual and thermal stimuli are capable of eliciting a thermoregulatory response. Negative orientation of locusts at midday may be modified by wind which affects body surface temperature (Waloff, 1963). In the formation of ground groups, locusts find and occupy the warmest spot on a laboratory cage floor, implying tarsal sensitivity to heat (Chapman, 1955). Both phototactic and photokinetic responses are involved in grouping responses (Cassier, 1965). Fraenkel and Gunn (1961) suggested that the thermoregulatory behaviour of locusts in the field contains elements of both taxis and kinesis.

ALTERATION OF FLIGHT BEHAVIOUR

Behavioural thermoregulation during continuous flight could be achieved by alternating between flapping and gliding flight. Insects having low wing loading, such as butterflies, dragonflies, and locusts, exhibit the capacity to glide for considerable periods of time. In dragonflies, most species glide for longer durations at high ambient temperature e.g. the percentage of time spent gliding by the dragonfly *Tramea carotina* increased linearly from about 20 to 45% as ambient temperature increased between 20 and 35° (May, 1978). Similarly, gliding is common in locust swarms. At air temperatures about 40°C upto 90% of the locusts in immature swarms glided for periods of upto 45 sec.

Under windy conditions, adult migrating locusts can glide almost continuously. Increased thoracic temperature due to heat production associated with the flight effort, has been suggested as the mechanism for the shift from flapping flight to soaring in migrating locusts (Uvarov, 1977), implying a thermoregulatory mechanism. Obviously a shift from powered flight to gliding flight will result in a reduction in endogenous heat production and increased durations of gliding at high ambient temperature are consistent with the hypothesis of thermoregulation.

In addition to gliding, migrating locusts may fly at a variety of altitudes. The decrease in temperature with altitude should allow flying locusts to ``select'' the appropriate ambient temperature that allows thoracic temperature to fall within the range optimal for muscle function. It has been suggested that the ability to fly at different altitudes has important consequences for maintaining water balance during migratory flights. A similar rationale could be applied for thermoregulation. However, little evidence is available to suggest that locusts actively control their altitude during flight (see Uvarov, 1977).

12

PHOTOPERIODISM AND INSECT BEHAVIOUR

Various aspects of insect behaviour as regulated by photoperiod have been describved in great detain by Beck (1968) and others. In the following pages only a brief account of the photoperiodism as associated with insect behaviour is given.

INSECT BEHAVIOUR IN RESPONSE TO DIURNAL RHYTHMS

Majority of the insect species clearly show daily behavioural activities that are associated with the locomotion, feeding, mating, oviposition and that may occur at typical times of the day. As elsewhere stated, many are active during day light hours (diurnal species), others tend to be active at night (nocturnal species), while still others are active mainly during evening or morning twilight (crepuscular species). These different activity habits are governed by daily cycles of temperature, humidity and light intensity. A nocturnal insect might be thought to be active only at night because its motor activity was inhibited by day light and if kept in continuous darkness its activity would be expected to be more or less continuous. Experimental evidences show that daily activity rhythms were at least partially endogenous and photoperiod playing a regulating role.

The experiments of Lutz (1932) under constant temperature and controlled photoperiod have indicated that the house cricket, *Acheta domesticus* and field cricket, *A. assimilis* become active shortly after the onset of darkness (scotophase) continued to move about actively for from 4 to 6 hours and were then relatively quiescent until the following day. When the crickets were held in constant darkness, they continued to display an activity rhythm in which a few hours of activity would occur once every 24 hours. If after a period of a few days of darkness, the insects were exposed to a reverse photoperiod, the activity rhythm adjusted to the new light dark schedule so that the time of maximum activity began shortly after the beginning of the scotophase, even though this occurred at noon. On the basis of his observations, Lutz (1932) concluded that the activity rhythm represented an endogenous periodicity that was subject to regulation by the environmental photoperiod. Observations of Lutz (1932) have been confirmed by a large number of workers with recent experiments.

It has been reported that in certain cases, the pupal moult in the fruitfly, *Dacus tryoni* occurred with a diurnal rhythm. The adult of *Drosophila melanogaster* emerged between the hours of 6.00 AM and 2.00 PM. when kept at constant temperature and normal sequence of day light and night. The rhythm of emergence persisted for some generations when the insects were reared in constant darkness (Andrewartha and Birch, 1954).

In insects, free running rhythms usually display periodicities that are only approximately 24 hours. The period of a rhythm is the time elasping from the beginning of one cycle to the beginning of next. This characteristic has been found in a wide variety of behavioural and physiological rhythms among organisms and is described by the term ``circadian" (meaning approximately daily) coined by Franz Halberg (Halberg *et al.,* 1959). The control of the behavioural rhythm to a reverse photoperiod may not be accomplished in one cycle of the rhythm but may result in the appearance of intermediate responses during a transition period of one to several cycles. These transitional responses are generally known as transients.

Under natural field conditions, insect activity is inflenced by several factors in addition to photoperiod and some of these factors may modify or totally suppress normally periodic behaviour. Temperature, humidity, light intensity as well as physiological factors such as age and reproductive state of the insect may strongly influence the manifestation of behavioural patterns e.g. young adults of the emperor dragon fly, *Anax imperator* tend to fly mainly during a few hours following dawn in preference to other times of the day. Older reproducing dragon flies do not display a dawn flight habit but tend to fly during mid day (Corbet, 1957, 1960). Age differences in daily patterns of flight, feeding and oviposition have been reported in a number of biting Diptera. The circadian activity rhythm of females of the cockroach, *Leucophaea maderae* was observed to vary with the reproductive state of the insect. During the time when mature eggs were present in the lower part of the reproductive tract, locomotor activity rhythms were greatly suppressed.

PHOTOPERIODIC CONTROL OF FEEDING RHYTHMS

As pointed out earlier, locomotor activity is usually accompanied by other behavioural manifestations such as feeding, mate seeking and oviposition. These different activities may, therefore, display a dial rhythmicity. The feeding activities of relatively sessile forms e.g. lepidopterous larvae, appear to form a reasonably simple rhythmic behaviour. On the other hand, the flight and feeding patterns of host seeking mosquitoes and other blood sucking Diptera are extremely complex.

Cockroaches tend to feed mainly at night which is also the time of their greatest locomotor activity. The possibility of the locomotor rhythm being influenced by a physiological hunger cycle was investigated by Harker (1956). Food was made available to the cockroaches, *Peripaneta americana,* only during the middle of the photophase. This affected the activity of insect during the hours when food was present but it did not exerted a phase setting influence on the general activity rhythm. The cockroaches continued to display maximum locomotor behaviour shortly after the beginning of the scotophase i.e. dark phase. They did not learn to anticipate the introduction of food during the photophase. This is in sharp contrast to the ability of worker honey bees, *Apis mellifera* to adjust their foraging activity in accord with previous experience and to display a learned circadian rhythm of feeding.

A daily pattern of feeding was observed in adults and nymphs of the house cricket, *Acheta domesticus*. A distinct circadian rhythm of feeding among nymphs of the milkweed bug, *Oncopeltus fasciatus* has been reported by Beck *et al.* (1958) but the role of photoperiod was not determined. Daily rhythms of general activity, mainly of feeding have been studied in several species of defoliating caterpillars. The feeding behaviour of mosquitoes and other blood sucking forms has been studied in detail. *Aedes aegypti* displays a diurnal biting rhythm, with the most active feeding periods occurring during the morning and just prior to sunset. *Aedes ingrami* showed a pre-sunset peak of feeding activitiy but *Aedes africanus* and other species studied were crepuscular. Moonlight did not affect the biting cycles of the mosquitoes. Corbet (1965) is of the view that the biting activity is a two or more step response to light intensities of a narrow range. The initial stimulation by light of the appropriate intensity would occur after sunset and again before sunrise.

PHOTOPERIODIC CONTROL OF MATING BEHAVIOUR

Very little experimental work has been done to see the effect of photoperiod on mating behaviour. Flitters (1964) studied the mating behaviour of the Mexican fruitfly and observed that the flies mated during late afternoon and early evening. When the photoperiod was epxerimentally reversed, the daily mating behaviour was also reversed so that the flies continued to mate during the latter part of the photophase. The existence of a photoperiodically controlled copulatory rhythm was suggested by these results. McCluskey (1958, 1963, 1965) studied the mating flight activity of a number of species of ants. Males of the harvester ant, *Veromessor andrei* entered a period of mating flight activity at the beginning of the daily photophase. The Argentine ant, *Iridomyrmex humilis* displayed such flight activity towards the end of the photophase, corresponding to a pre-sunset field behaviour. Males of the fire ant, *Solenopsis saevissima* showed mating flight behaviour during the middle and latter half of the photophase. *Componotus clarithorax* males tended to fly early in the photophase, corresponding to a dawn flight under field conditions. The flight behaviour of these species of ants was found to be an endogenous circadian rhythm under constant conditions. Mating flights by the males of the honey bee have also been reported to follow a circadian rhythmicity (Taber, 1964).

Rhythmicity in the sexual excitability of male moths as mediated by sex pheromones has been studied by Allen *et al.* (1962; Shorey, 1964; Jacobson, 1972). In several species of moths, it has been demonstrated that the rhythm of male responsiveness was shown to be controlled by photoperiod with maximum responsiveness occurring in the early scotophase i.e. dark phase and experimental alterations in the laboratory photoperiod caused corresponding phase adjustment in the male response rhythm.

INFLUENCE OF DIURNAL RHYTHM ON SEX PHEROMONE PRODUCTION AND MATING BEHAVIOUR

In the following pages a brief account of the photoperiodic influence of sex pheromone on mating behaviour has been given. (for details see : Jacobson, 1972).

Dictyoptera

Adult males of the *Periplaneta americana* exhibit daily rhythms in responding to the sex attractants of virgin female. Peak responses were noted at 2.00 AM and the lowest responses at 8.00 AM with normal photoperiod tests. In reversed photoperiod tests, peak responses were at 2.00 PM and the lowest at 8.00 PM.

Lepidoptera

This order comprises of many species, mainly moths in which mating normally occurs during the hours of darkness and many others mainly butterflies in which assembling and mating occur during day light hours. Tamaki *et al.* (1969) observed adult *Adoxophyes orana,* one to four days old mated about one hour before sunrise or between 2.00 and 4.00 AM. According to Rejesus and Reynolds (1970) the peak mating activities or presumably pheromone release by the cotton leaf perforator, *Bucculatrix thurberiella* occurred between 5.00 to 8.00 PM and a smaller peak activity occurred between 1.00 and 2.00 AM and the number of males attracted therafter decreased until no moths were caused at mid night.

Sanders (1971) recorded that the female *Choristoneura fumiferana* started calling earlier (by sunset) than female *C. pinus* i.e. about 1.5 hours after sunset. Male *C. fumiferana* are also attracted by females to traps earlier, although the times of the peak catches were approximately the same in both species (Sanders, 1971). Gentry *et al.* (1964) observed that the male *Heliothis virescens* are attracted to females only in the early morning hours between 4 AM and day light.

According to Ouye and Butt (1962) although male *Pectinophora gossypiella* are attracted to females in the field during darkness, those males reared in the laboratory under continuous illumination will respond readily at any time to an extract of the female abdomens. Barth (1937, 1938) reported that the production of sex attractant by females of *Plodia interpunctella, Anagasta kuehniella, Achroia grisella* and *Galleria mellonella* does not appear to be restricted to specific hours. Traynier (1970) stated that virgin female *A. kuehniella* emit the sex pheromone at the beginning of the light period of a 24 hour cycle of light and dark periods. The response of males was found to be maximal at dawn. Thus, it is clear that lepidopterous insects exhibit specific but extremely varied times of courtship and mating.

Hymenoptera

It has been found that the attraction of male *Bracon hebetor* by females occurs during day light i.e. morning hours. A virgin female *Diprion similis* set in the field at 11 AM attracted its first male within 30 seconds. Activity continued until 4 PM, when a total of over 7000 males had been attracted. The greatest activity occurred between midnight and midafternoon as reported by Coppel *et al.* (1960).

Diptera

In *Dacus tryoni*, mating is normally confined to the dusk period. Responses of females to the male pheromone were observed only in the late after noon and dusk period. The greatest activity of female hessain flies, *Phytophaga destructor* to attract males occurs in the early morning.

Coleoptera

It has been reported by Gara (1963) and Vite *et al.* (1963) that the volatile substance responsible for mass attraction of both sexes to male *Ips confusus* is produced only during the initial period of attack on ponderosa pine logs, but the attraction shows a definite diurnal pattern of 8 to 10 AM and 2 to 6 PM. The male *Xenorhipis brendeli* shows its greatest responses to females between 12.30 and 3.00 PM daily with more than 80% of all males being captured between 12.40 and 1.30 PM (see : Jacobson, 1972).

REVERSAL OF REPRODUCTIVE DIAPAUSE BY PHOTOPERIOD AND TEMPERATURE IN GRASSHOPPERS

Photoperiod and temperature play a major role in the regulation of insect diapause. The grasshopper, *Oedipoda miniata* exhibits a reproductive diapause during the hot and dry summer (May-October) in Israel. The diapausing adults feed well and are active but females do not lay eggs and males show only extremely slight mating behaviour (Pener and Broza, 1971; Broza and Pener, 1972). In the laboratory, photoperiod and temperatures correspon-ding to autumn in Israel terminated diapause and induced intense male sexual behaviour (Broza and Pener, 1972). Intense mating behaviour was also induced by short (LD 10.5 : 13.5) or medium LD 12 : 12) day length at relatively low (26° - 13°C) or high (38° - 27°C temperatures. Continuous illumination (LL) and high but inconstant temperatures maintained diapause in both the sexes and even reinduced it in field born post-diapause adults (Pener and Broza, 1971; Broza and Pener, 1972).

Observations of Orshan and Pener (1979a) have indicated that short day length and relatively low temperatures (LD 10.5 : 13.5; 26° during illumination and 13° during darkness) induced mating behaviour and terminated aestival reproductive diapause in males of the grasshopper, *Oedipoda miniata*. Long days and high temperatures [LD 16 : 8, 38° during illumination (photophase) and 27° during darkness (scotophase)] maintained sexual inactivity and diapause. Repeated changes of these conditions terminated, reinduced, reterminated and reinduced, again diapause in the same males. An endogenous trend to complete the diapause development and terminate diapause does not seem to exist in the males of this grasshopper, their reproductive diapause may be regarded as rather primitive.

It has been further remarked by Pener and Broza (1971) and Broza and Pener (1972) that during the warm and dry summer in Israel, adults of *O. miniata* exhibit a reproductive aestivation. Depending on spring temperatures in April to May and related to geographical locality (hills or low lands) moult to adult stage occurs in May-June. The grasshoppers feed and are quite active but females do not lay eggs and males show very little mating behaviour. This state lasts till the end of September to the beginning of October when reproductive diapause is terminated in the field. The termination and reinduction of reproductive diapause as related to photoperiod and temperature in males of *O. miniata* has been described in considerable detail by Orshan and Pener (1979 a, b).

ROLE OF PHOTOPERIOD ON THE INCIDENCE AND PHYSIOLOGY OF DIAPAUSE IN LEPIDOPTERA

Mansingh and Smallman (1967) studied the effect of photoperiod on the incidence and physiology of diapause in Lepidoptera and remarked that the obligatory diapause in the univoltine saturniid species, *Hyalophora cecropia* and *Antheraea polyphemus* was found to be controlled by photoperiod. Diapause in both the species was induced and maintained by short day photoperiods of 8 to 12 hours and prevented or terminated by long day photoperiods of 17 to 24 hours. The incidence of diapause depended mostly on the light regimes imposed on fourth or fifth instar larvae. Wide individual variations were observed in the intensity and duration of diapause.

PHOTOPERIODIC CONTROL OF OVIPOSITION

A number of workers (Haddow, 1954; Gillett and Haddow, 1957, Haddow and Gillett, 1957) have suggested that the rhythm of biting activity in mosquitoes is closely related to their ovipositional rhythm. The ovipositional behaviour of *Aedes aegypti* is a photoperiodically controlled rhythm was clearly demonstrated by Haddow and Gillett (1957). Under natural conditions, oviposition occurred mainly between 2 and 3 PM. Under a controlled laboratory photoperiod, daily ovipositional activity was concentrated in the first hour of the scotophase even when the photoperiod has been experimentally modified or reversed. *Aedes apicoargenteus* and *A. africanus* were also shown to oviposit on a rhythmic basis (Haddow *et al.*, 1960).

13

PHOTOPERIODISM AND ECOLOGICAL ADAPTATIONS

Photoperiodism plays an important role in the complex ecological adaptations of insects. As an environmental factor, photoperiod differs from physical forces such as temperature, moisture, radiation, air current as well as pressure and also from chemical factors as represented by nutritious and harmful substances that are found in the environment. These several physical and chemical factors may pose either threats or opportunities for the continued survival of species by their direct action on the biological system. Photoperiod, however, exerts no directly beneficial or harmful effects on the organisms but provides an adaptative opportunity as a source of temporal information. Its dawn and dusk signals and the time interval between them are available for utilization as informational inputs from the environment. The adaptations of insects to photoperiod have evolved as ordered responses to these stimuli. Photoperiodic determination of polymorphic forms of aphids for example represents an important adaptive response. The polymorphism in aphids represents synchrony between the biology of aphids, the biology of the host plants of the aphids and the seasonal climatic cycle of the environment, rather than in any direct physical effects that day light and darkness might be postulated to have exerted on the aphid itself (Beck, 1968).

Photoperiodic adaptations involve the reception of light on and light off stimuli as an input of bits of environmental information. These token stimuli influence the internal temporal organization of the insect by virtue of their phase setting effects on physiological rhythms. Genetically determined behavioural and developmental alternatives are probably implemented in accord with the temporal organization of the living system. Thus, the responses of insect to photoperiod may become an integral part of from one to several of the organisms ecological adaptations. Beck (1968) has given a comprehensive account on various aspects of photoperiodic control of ecological adaptations e.g. conservation of water, acquisition of food, reproduction, seasonal development, short day and long day adaptations, aestivation and hibernation and the importance of photoperiodism in univolatine insects. He has also described the influenced of photoperiodism on geographical populations, adult emergence rhythms, induction, termination and determination of different types of diapauses in considerable detail.

PHOTOPERIOD AND BEHAVIOUR OF DESERT INSECTS

Among the desert orthopteroid insects, one enconters two broad categories of light associated diel behaviour. A diurnal pattern applies to many herbivores such as locusts, grasshoppers and phasmids as well as to carnivores like mantids. Many of these insects fly, hop or wander actively in search of their hosts. However, phasmids tend to be sedentary. Studies conducted on the migratory locusts suggest an indirect explanation for the relatively weak diurnal circadian rhythms of *Locusta migratoria* (Edney, 1937) and *Schistocerca gregaria* (Cloudsley-Thompson, 1977). The flexibility of their responses may allow these species to take advantage of ephemerally favourable food supplies and oviposition substrates, whereas if their rhythms were well entrained by photoperiod, this opportunistic behaviour would not be possible (Cloudsley-Thompson, 1977). While the locusts and grasshoppers show diurnal behaviour, a number of other orthopteroids seem to be nocturnal. This category of light associated behaviour characterizes many fossorial and crevice inhabiting detritivores including gryllids, gryllacrids and cockroaches.

Opportunistic use of photoperiod, moisture and temperature by the Australian plague locust, *Chortoicetes terminifera* is reviewed by Matthews (1976). This arid zone species lays diapause as well as non-diapause eggs. The latter are inserted more deeply into the soil in summer than in winter, thereby avoiding an inappropriate hatching stimulus of light rains arriving during hot weather. Similar rains promote hatching of non-diapause eggs near the surface in winter but diapause eggs hatch later. Thus, an array of hatching possibilities, conditioned by a combination of genetic and environmental factors is available to this species of locust and helps to ensure its survival in an unpredictable habitat.

REVERSAL OF PHOTOTAXES ASSOCIATED WITH ADAPTATION

The potato beetle, *Leptinotarsa decemlineata* is normally photo-positive but it may become photonegative as it becomes desiccated. During dry weather, the beetles move down the stems of plants into the soil where they remain until they have been remoistened by rain. They then become photopositive again and ascend the stems of plants where they lay their eggs.

Reversal of phototaxis at high temperatures is also shown by larvae of *Malacosoma* and *Neodiprion*. The tse-tse fly, *Glossina morsitans* is positively phototactic at normal temperatures but at temperatures of 32°C and above, it becomes negatively phototactic. Adults of the syrphid, *Eristalis tenax* move towards light at temperatures between 10° and 30°C but above 30°C they move away from the light (i.e. becomes negatively phototactic).

In some insects a reversal of phototaxis occurs as they grow older. This is common among blow flies. Larvae of *Calliphora erythrocephala* are photo negative all their lives but the intensity of the response varies with age, reaching a maximum about the 4th day of the larval life after which it falls until the 7th day when it remains constant until pupation. Between 4th and 5th days, about 15% of the larvae migrate to pupate, by the 6th day 60% migrate and by the 7th day, practically all would have migrated. The drop in the intensity of negative phototaxis thus coincided with migration. The reversal of the phototaxis as observed in several instances has distict adaptive value (Andrewartha and Birch, 1954).

INSECT BEHAVIOUR IN RESPONSE TO INTENSITY OF LIGHT

It has been observed that the bug, *Notonecta* when moving in the presence of a discrete source of light will tend to keep the long and transverse axes of its body perpendicular to the path of light. This is called the *dorsal light reaction*. Other species may so direct their movements temporarily at least that their path makes a constant angle with the path of the light. This sort of behaviour is called ``light compassing movements''. When the light is very distant, the insect traces a straight path but when light is close e.g. an electrical bulb in a room, the path is curved. Light compassing movements are especially important in ants and bees in their daily excursions after food (Fraenkel and Gunn, 1940).

Some insects respond the direction of the light rays. Telotaxis appears to be of this nature. Similarly ``light compassing movements'' are movements in relation to direction of light rays and not to the intensity of light. It is certain that light compassing movements involve the formation of an image of the source of light and photokinesis depends on comparison of the intensity of light coming from different directions with respect to the position of the body. Fraenkel and Gunn (1940) have stated that in taxes photopositive animals, in this case, insects, go towards the light source even if the intensity is lower there and photonegative ones, go away from the source even if they reach a higher intensity, it looks as if intensity is unimportant.

Thus, photoperiodism and/or intensity of light both are closely associated with behavioural, ecological and other types of adaptations of insects (for details see : Beck, 1968).

14

HORMONAL CONTROL OF INSECT BEHAVIOUR

The insect behaviour is greatly influenced mainly by the neurohormone of the brain, juvenile hormone of the corpora allata and ecdysone of the ecdysial glands. Along with the "kinds of insect behaviour" only a brief account of the endocrine (hormonal) control of insect behaviour was given, and in the following pages, the hormonal control of insect behaviour has been dealt with some detail, which is based on a number of research papers and the review articles.

HORMONAL CONTROL OF FLIGHT

Experimental studies carried out on locusts and some other insects show necessity of certain neurohormones for flight to occur. Removal of the neurosecretory cells of the pars intercerebralis i.e. medial neurosecretory cells, causes the disappearance of nocturnal flight in the noctuid moths, *Noctua pronula* and *Agrotis ipsilon*. In the blow fly, *Calliphora erythrocephala*, the corpora cardiaca are essential to the achievement of flight. In the locust, *Schistocerca gregaria*, the neurohaemal part of the corpora cardiaca performs an important role in sustained flight, which is prevented by removal of the corpora cardiaca and is only restored after their reimplantation. According to Michel and Bernard (1973), flight in *S. gregaria* is stimulated by a substance produced in the pars intercerebralis (the brain hormone) and activated in the corpora cardiaca.

HORMONAL CONTROL OF CIRCADIAN RHYTHMS

Studies of numerous endocrinologists have shown that hormones play a key role in daily activity cycle of the insects in general. It is presumed that a biological clock controlled circadian rhythm is located in the brain, most probably near to the pars intercerebralis, because various brain injuries or extensive removal of pars intercerebralis causes the circadian rhythm to stop. The endocrine regulation of circadian rhythm hypothesis is supported by the demonstration of a circadian activity rhythm in the neurosecretory cells of the pars intercerebralis, as well as in the A-type of neurosecretory cells of the ganglia of the ventral nerve cord. Experiments in which the implantation of a gel containing actinomycin-D, a protein synthesis inhibitor, when introduced in the pars intercerebralis makes the *Leucophaea maderae* arrhythmic.

In the house cricket, *Acheta domesticus,* removal of the pars intercerebralis upsets the locomotor rhythm and causes hyperactivity. Experimental work has shown that there is a cyclic serotonin secretion in the corpus centrale that controls the activity of the medial neurosecretory A - cells of the brain which in turn inhibits insect activity.

Experimental evidences in hand have clearly shown that either the brain, particularly medial neurosecretory cells or the A-cells of suboesophageal ganglion control the activity rhythm. In *Teleogryllus commodus* removal of pars intercerebralis medialis of the protocerebrum makes the crickets arrhythmic. In the grasshopper, *Romalea microptera,* brain removal allows the locomotor rhythm to continue but the latter is prevented by removal of the suboesophageal ganglion. Investigations carried out by Dupont-Raabe (1956) and Eidmann (1956) in the stick insect, *Carausius morosus,* which displays circadian colour change and circadian activity rhythm both with a nocturnal activity period. It was found that the triggering mechanisms of these two processes i.e. circadian colour change and circadian activity rhythm during night hours are different; whereas neuroendocrine control occurs in the former case, complete nervous control occurs in the latter case. Removal of the corpora cardiaca - corpora allata complex and the pars intercerebralis has no effect on behaviour. However, removal of the brain or cutting of the circum oesophageal connective leads to the disappearance of the circadian rhythm and increased activity showing that rhythm control is exerted by the brain and is transmitted nervously. Removal of the sub-oesophageal ganglion or cutting the connectives behind this organ permanently immobilizes the insects. The cutting of the ventral nerve cord behind the sub-oesophageal ganglion allows circadian activity to continue in the anterior part of the body whereas the posterior part remains motionless (Dupont-Raabe, 1956).

HORMONAL CONTROL OF ECLOSION RHYTHM

It has been normally observed that several species of insects e.g. *Hyalophora cecropia* and *Antheraea pernyi,* always emerge at a definite time which varies for each species. Eclosion regulation is under the influence of brain as is indicated by removal and implantation experiments. Thus brain plays a prominent role. The active factor termed ``eclosion hormone" is secreted by the neurosecretory cells of the pars intercerebralis (Truman, 1973) which is stored in the corpora cardiaca. It triggers eclosion behaviour. Detailed process of eclosion, controlled by eclosion hormone has been described by Truman (1973). The time of appearance of eclosion hormone in the haemolymph of *Manduca sexta* was determined by isolated wing bioassay. It was demonstrated that this hormone is released in the blood only at a restricted time which is determined by a circadian clock once development has completed.

Eclosion hormone is present in several lepidopterous insects e.g. *Hyalophora cecropia, Antheraea pernyi, Manduca sexta, Bombyx mori* etc. The tanning processes of the cuticle begin as soon as the young moth has spread its wings. The question arises whether bursicon i.e. tanning hormone is different from eclosion hormone. By using different organs of *Manduca sexta,* Truman (1973) tested eclosion hormone activity on *Antheraea* and bursicon activity on *Manduca* (Truman, 1973). These experiments performed by Truman revealed that eclosion hormone is located in the brain and in the corpora cardiaca, whereas bursicon is found in the abdominal ganglia and in the perisympathetic organs. Removal of different parts of the brain has indicated that eclosion hormone is located in the pars intercerebralis medialis of the protocerebrum and not in the lateral protocerebral neurosecretory cells.

HORMONAL CONTROL OF MATING BEHAVIOUR

Mating in insects is often initiated by the production of pheromones, which allow the male and female to meet (see : Reproductive behaviour). Other compounds originating from food or from the offspring host plants also play an important role in the meeting of the partners as well as the production of light and sound. Hormonal regulation of pheromone production and sexual behaviour both involve juvenile hormone but not in the exactly same way. As far as neurohormones are concerned, their involvement has been shown in the following insects.

In cockroaches, female sex pheromone production depends on the corpus allatum (Barth, 1962, Roth and Barth, 1964) but the sexual receptivity of female is only partly controlled by these organs in *Nauphoeta, Byrsotria, Leucophaea* and *Blaberus* (Roth and Barth, 1964). The possibility that the pars intercerebralis medialis of the protocerebrum helps to control mating behaviour was suggested by Barth (1962) and Engelmann (1970).

Barth in a series of papers (1962, 1963, 1965) has shown that CA plays a key role in mating behaviour of the cockroach, *Byrsotria fumigata.* Its virgin females were found to produce sex pheromone. The production of this attractant is controlled by CA-hormone (JH). He further suggested that CA are the intermediate link in the control of sex pheromone production and other pheromone (s) associated with reproductive cycle. Pener (1967) has also pointed out that the CA control sexual behaviour and colour change in *S. gregaria.*

As in the case of cockroaches, the production of pheromone and sexual behaviour of locusts depend in varying degree on the corpora allata. In *Schistocerca* as in *Nomadacris* male deprived of their corpora allata were not able to copulate. The same was observed in *Locusta.* Recent work including precocene application and azadirac-htin injection to young larvae of *Locusta* revealed that the over aged insects resulting from these treatments exhibited sexual behaviour despite their degenerated corpora allata. It has been shown that out of A, B and C cells, of the brains, the C cells only control sexual behaviour. A number of workers have clearly shown that the mating behaviour in some insects not all, is definitely controlled by the hormones of the brain and corpus allatum.

In some species of lepidopterous insects also, hormone controls mating behaviour e.g. sex pheromone production in *Heliothis* is controlled by a neurohormone from the brain which is present in the haemolymph during the scotophase only. The extractible pheromone content of the female was strongly reduced after neck ligation but increased again when ligated females were injected with brain extracts. But in several other lepidopterans calling behaviour depends on neural control.

The sexual behaviour of some male dipterous insects depends on the presence of corpora allata and the female sexual receptivity can be induced by corpus allatum implantation or by the application of juvenile hormone or its analogue (Raabe, 1989).

BEHAVIOUR OF COCOON CONSTRUCTION

The behaviour of cocoon construction in Lepidoptera is also under the influence of hormones. It has been found that for the construction of cocoon a low level of juvenile hormone is needed but the type of cocoon constructed is controlled by the neurosecretory cells of the brain. In *Samia cynthia* a hormone from the neurosecretory cells of the brain directs the construction of either summer or a winter cocoon (Truman and Riddiford, 1974).

The cocoon construction behaviour was greatly influenced by the application of juvenile hormone analogue. The juvenile hormone analogue treated larvae or prepupae of *Philosamia recini* survived for 24 days and died without metamorphosis, majority of them failed to spin cocoons and formed naked pupae; cocoons if formed were extremely thin through which prepupae or pupae were clearly seen. They were light in weight with 2-big valves (Awasthi, 1985).

HORMONAL CONTROL OF REPRODUCTIVE BEHAVIOUR

As with other groups of animals, the reproductive behaviour of insects has received considerable attention. The behaviour of the female is usually complex in that it may markedly change a number of times during her life. The newly emerged female is often sexually immature and unreceptive to males. During the following days sexual behaviour then appears in concert with, but independent of, the maturation of the ovaries. After courtship and successful copulation, her behaviour changes; receptive behaviour disappears and her activities are directed towards the deposition of eggs.

The onset of sexual receptivity in virgin females has most often been linked to juvenile hormone. The first evidence for such a role came from studies on the cockroach, *Leucophaea maderae* that showed that removal of the corpora allata from newly ecdysed females drastically reduced the percentage of females that subsequently became sexually receptive. Similarly, in the grasshopper *Gomphocerus rufus* allatectomy of newly emerged females completely prevented the onset of receptivity. They remained in a permanent state of "primary defence" and consistently rejected courting males. Implantation of active corpora allata or injection of juvenile hormone mimics into such females subsequently resulted in the development of ``copulatory readiness". In the Diptera *Drosophila melanogaster, Musca domestica* and *Aedes aegypti* and some Hemiptera, juvenile hormone also seems to promote the maturation of sexual behaviour.

Juvenile hormone may have a number of effects in the adult female; it is usually required for maturation of the ovaries and is often necessary for the production of accessory gland secretions and of sex pheromone. Because of these other actions, its role in sexual behaviour can sometimes be confused. A case in point is Barth's (1961, 1962) study in mating behaviour in the cockroach *Byrsotria fumigata,* a species in which sex pheromone production is under juvenile hormone control. Allatectomized virgin female *Byrsotria* did not mate if placed with males. However, this result was due not to their failure to become sexually receptive, but rather to the fact that they did not produce the volatile pheromone needed to attract the male. When such females were painted with pheromone extract, they were then courted vigorously and mated readily.

In some grasshoppers and cockroaches a neurosecretion from the brain is also implicated in the onset of sexual receptivity. In many instances, it is required after juvenile hormone has presumably modified the nervous system such that it can respond to the neurohormone. In the wild silk moths, juvenile hormone is not needed for sexual behaviour but instead a hormone from the corpora cardiaca serves to trigger pheromone release behaviour in the virgin female.

In a number of species a neurohormone from the brain or from the corpora cardiaca of the female has been implicated in ovipoistion behaviour. Presumably it acts on the central nervous system to stimulate the neural pathways involved in egg deposition. But the only neurophysiological demonstration of increased spontaneous motor bursts in the nerves leading to the ovipositor after mating has been in the silk moth *Bombyx mori,* in which no hormonal control has yet been shown. This neurosecretion also appears to have a direct myotropic action on the ovarioles, which would allow coordination between the movement of eggs from the ovaries and the behaviour involved in their deposition in a suitable site.

The males of most insects show a much simpler behavioural repertoire than do the females. Mate-finding, courtship, and copulation are the main activities of the male, and in most species there is no evidence that hormones are required for the maturation of this male behaviour. At or shortly after adult ecdysis the male shows full behaviour when confronted with the proper stimulus, and this condition persists throughout his life.

HORMONAL CONTROL OF EGG CARE

The dermapteran, *Labidura riparia* has a very unusual behaviour related to pre and post oviposition. During the 10 days of embryonic development, the female takes care of its eggs, turning and licking them does not eat and acquires aggressive behaviour. Pars intercerebralis cauterization switches off this behaviour, thus suggesting a neurohormonal control (Caussanel *et al.,* 1978). Ultrastructural study of the aortic wall during the reproductive cycle shows that whereas one peak of exocytotic profiles occurs at oviposition, a continuous exocytotic activity takes place during the egg-care period. (Juberthie and Caussanel, 1980).

BEHAVIOUR DURING GROWTH AND DEVELOPMENT

Besides their roles in the reproductive behaviour of the adult, hormones also regulate various behaviours during earlier stages of the insect life history. During larval life, the periodic moults that produce the cuticle for the next instar are often accompanied by a reduction in locomotor activity. There is evidence from larval locusts that this quiescence is due to the high titers of ecdysone present in the blood at this time.

In insects having complete metamorphosis, the full-grown larva performs a sequence of pre-metamorphic behaviours in preparation for pupation. In lepidopteran larvae this is marked by a voiding of the gut an increase in locomotor activity, and the construction of a silken cocoon or a pupation chamber. The absence of juvenile hormone is a prerequisite for the onset of cocoon spinning behaviour. The removal of the corpora allata from young larvae results in the cessation of feeding and the precocious construction of a miniature cocoon. Likewise, application of juvenile hormone late in larval life serves to delay cocoon spinning or to force spinning of a moulting pad characteristic of earlier larval instars. Although a low juvenile hormone titer is required for the display of premetamorphic behaviours, in and of itself it will not cause the behaviours to occur. They are apparently triggered through the action of the brain prothoracic gland axis in an environment of low juvenile hormone (Truman and Riddiford, 1974). The end of adult development in holometabolous insects is signalled by the emergence of the insect from the old pupal skin. In the moths, this emergence behaviour as well as the activation of adult behaviour patterns that accompanies it, occur in response to a peptide hormone (the eclosion hormone) that is released from the brain of the moth (Truman, 1973).

HORMONAL CONTROL OF POLYMORPHISM

Polymorphism is a widespread phenomenon in insects and has many determining factors and aspects, including morphology and pigmentation of the insect, its reproductive mode and its behavioural characters. On the whole, polymorphism allows specialization which facilitates the dispersion of the species and above all the sharing of work in social insects.

The role of corpora allata has been established in triggering of green pigmentation, wing development, rate of maturation of ovaries, certain behavioural characteristics of solitary phase in locusts, the phase determination of their offspring, and bee and termite caste determination.

In the lepidopterans, *Polygonia c-aureum, Papilio xuthus* and *Lycaena phalaeas,* which may develop in distinct spring or summer morphs, a hormonal role of the brain has been shown. It has been also reported that the corpora allata and the neurohormone of the brain are involved in morph determination in the aphid, *Megoura viciae*.

It can be concluded that the polymorphism in insects is controlled by the juvenile hormone and the neurohormone of the brain and both are equally important.

15

PHEROMONAL CONTROL OF INSECT BEHAVIOUR

The social attraction of an insect colony appears to be maintained primarily through the utilization of a series of communication chemicals which function to transmit specific bits of information to the members of a single population. These chemicals are termed as "pheromones". The term pheromone was coined by Karlson and Butenandt (1959). According to Karlson and Butenandt (1959) the pheromones are chemicals which are secreted to the outside by animals, in this case insects, and which if passed to another individual of the same species, cause it to respond in a particular manner. These are produced in several specialized tissues or glands called exocrine glands, which evacuate their contents into the environment in response to specific stimuli. Social insects such as hymenopterans and isopterans, possess many types of exocrine glands. It is well known that the pheromones modifies the insect behaviour in a variety of ways and now several hundred pheromones are known whose chemical properties etc. have been fully elucidated (see : Beroza, 1970 and Jacobson, 1972). Many aspects of insect biology and behaviour are controlled by these chemicals - the pheromones. The pheromones which have been identified are sex pheromones, trail marking pheromones, swarming pheromones, alarm pheromones, aggregation pheromones, territorial - marking pheromones and spacing pheromones etc. Account of some of these pheromones is given in the following pages.

The insect pheromones have been reported from Orthoptera, Heteroptera, Homoptera, Diptera, Isoptera, Neuroptera, Siphonaptera, Coleoptera, Hymenoptera and Lepidoptera. In some of the cases, they have been studied in great detail e.g. Hymenoptera,

Orthoptera, Diptera, Coleoptera and Lepidoptera while in other groups comparatively less information is available. The pheromones coordinate and control a variety of biological and behavioural physiology in insects.

SEX PHEROMONES

Male and female conspecific insects often communicate with chemical *sex pheromones*. Mate location and courtship may involve chemicals in two stages, with *sex attraction pheromones* acting at a distance followed by closeup *courtship pheromones* used prior to mating. The sex pheromones involved in attraction often differ from those used in courtship. Production and release of sex attractant pheromones, tend to be restricted to one sex, mostly the female insects (Fig. 60 A,B) although there are several lepidopterans and mecopterans in which males are the releasers of distance attractants that excite females. The producer releases volatile pheromones that stimulate characteristic behaviour in those members of the opposite sex within the range of the odorous plume. An stimulated recipient raises the antennae, orientates towards the source and walks or flies upwind often in a zig-zag track until the source is reached. Each successive action appears to depend upon an increase in concentration of this air borne pheromone. As the insect approaches the source, cues such as sound and vision may be involved in close-up courtship behaviour (Gullan and Cranston, 1994).

Fig. 60A. A female sesiid moth releasing calling pheromone from everted gland at the abdominal tip.

Fig. 60B. This worker honey bee is exposing scent glands on her abdomen (Arrow) and fanning her own scent into the air with her wings

It has been observed that the courtship which involves coordination of the two sexes may require close-up chemical stimulation of the partner with courtship pheromone. This pheromone may be simply a high concentration of the attractant pheromone but *aphrodisiac* chemicals exist as seen in the butterfly *Danaus gilippus*. The males of this species as with several other lepidopterans have *extrusible abdominal hair pencils*, which produce a pheromone that is dusted directly on to the antennae of the female while both are in flight. The effect of this pheromone is to placate the natural escape reaction of the female who descends, folds her wings, and allows copulation. In *D. gilippus* this male courtship pheromone, a pyrrolizidine alkaloid is called "danaidone". The scent pencils are also present in Heliconian butterfly that secrete and disperse attractant or aphrodisiac pheromone (Fig. 60C). It is of profound importance to note that the butterfly cannot synthesize it without acquiring the chemical precursor by feeding on selected plants as an adult. Likewise in several butterflies and moths, the precursor is acquired through feeding the specific plant species. As remarked earlier, sex pheromones are mostly species specific and they are the mixture of several chemicals. Therefore, a single component of the mixture may elicit only one part of the sex attraction behaviour or a partial or complete mixture may be needed. Often the mixture produces a greater response than any individual component, a synergism which is widespread in insects that produce pheromone mixtures.

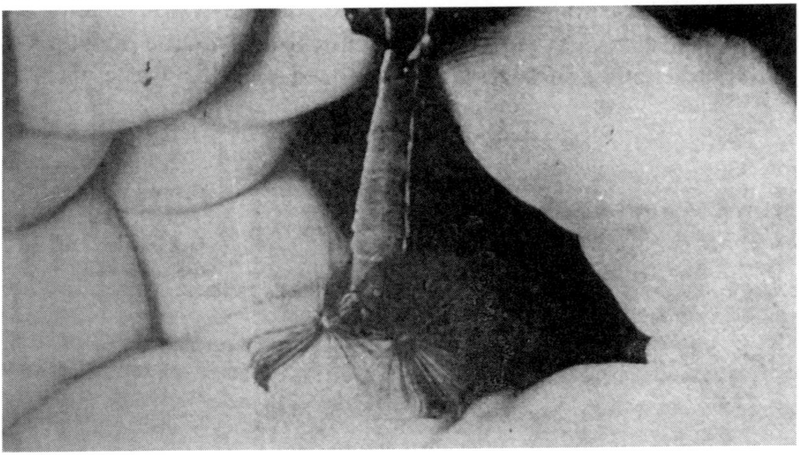

Fig. 60C. Scent pencils on male Heliconian butterfly from Ecuador

In certain moths, the virgin female assumes a calling position and emits from glands near the tip of the abdomen (Fig. 60A), specific *sex attractant pheromones* that diffuse out into the environment. These pheromones are usually specific and are detected by antennal receptors located on the male antennae (Fig. 61 A,B). If a minimal threshold of molecules (10^{-12} μg for the gypsy moth) come in contact with the antennal receptors, the male becomes "excited" and flies upwind, the most likely direction of pheromone source. Olfactory attractants bring the male close to the female, where other sensory stimuli such as: visual, tactile, or wing beat clues become important. The distance of initial attraction varies. Male gypsy moths have been reported to fly nearly 4 km to a female, but effective attraction is probably less than 100 m (Elzinga, 1978). It was also found that the males of the Indian moon moth *Actias selene* (Fig. 61C) liberated at a distance of 11 km., were able to return to the females (Wigglesworth, 1965).

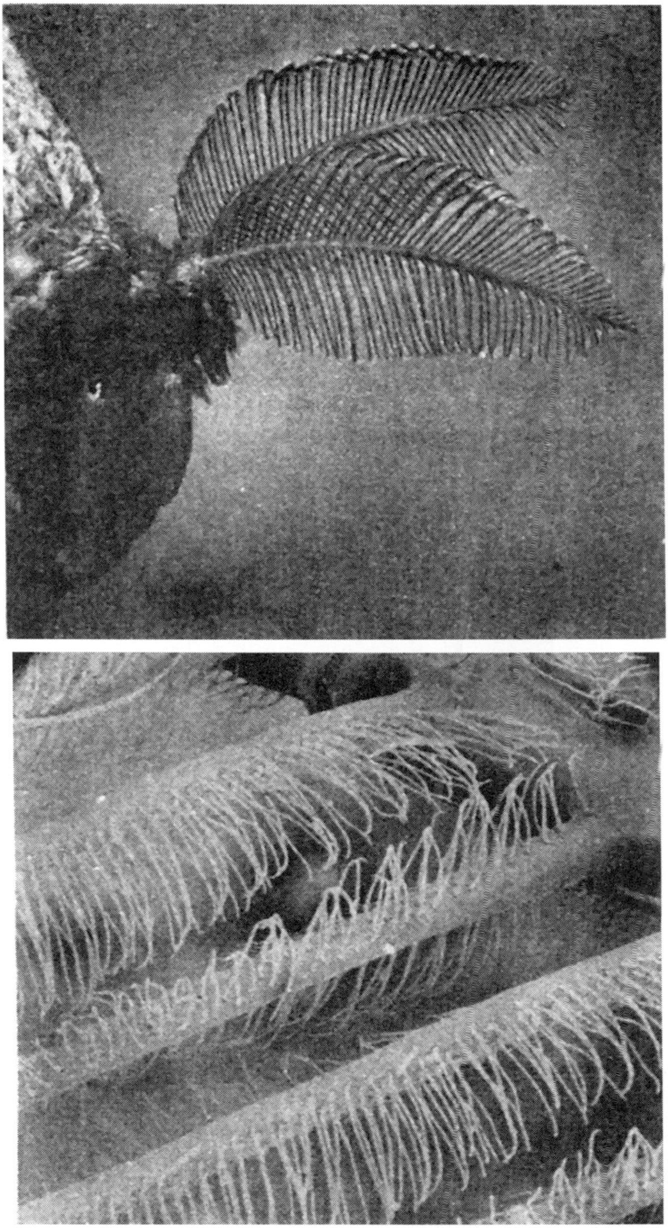

A

B

Fig. 61 A,B. Pectinate antennae of giant silkworm moth (*Antheraea polyphemus*) (A) lateral view; (B) SEM micrograph of filaments showing sensory setae used in detecting female sex pheromones

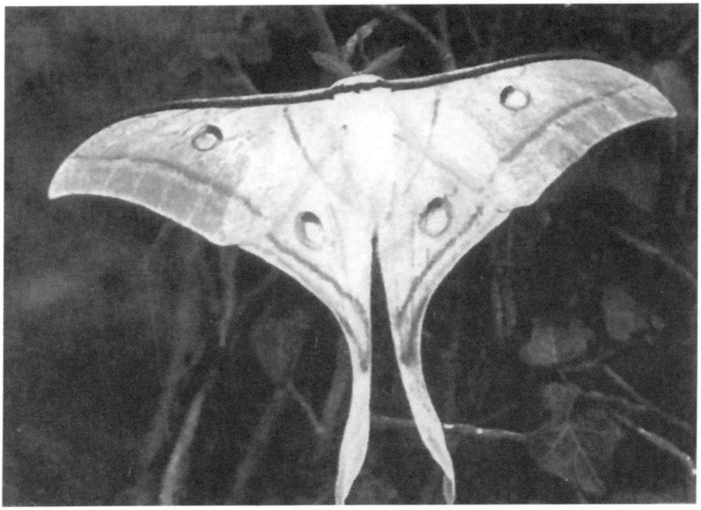

Fig. 61C. Indian moon moth, *Actias selene*

TRAIL MARKING PHEROMONES

Many social insects use pheromones to mark their trails particularly to food and nest. Most of the trail marking pheromones are volatile and short lived chemicals that evaporate within days unless reinforced. As stated above, the trail pheromones are found in insects of the order Hymenoptera and Isoptera (Fig. 62 A,B).

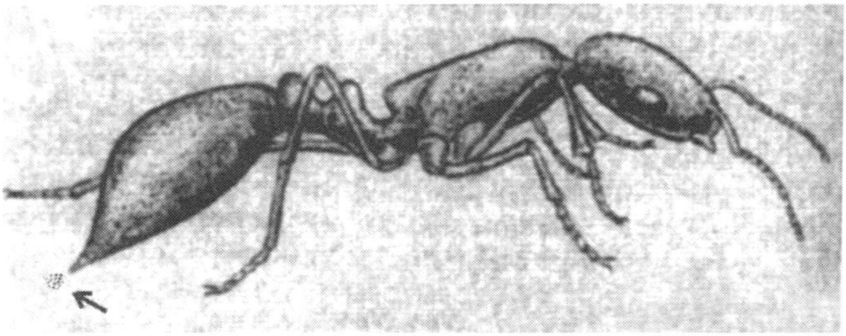

Fig. 62A. An ant leaving a scent trail which leads other ants to food objects (Arrow)

Ant Trail Pheromones

The ant trail pheromones originate in several different glandular tissues like hindgut, mandibular gland, poison gland, Dufour's gland, Pavan's gland and metathoracic legs of formicidae (Hymenoptera). Odour trail pheromones apparently originate in the hindgut in three sub-families of ants but the exact place of origin of the hindgut trail pheromone is not clearly understood. The Dolichoderine ants have developed a novel social organ, the Pavan's gland in which the trail pheromones probably are synthesized (Wilson and Pavan, 1959). Since this exocrine structure is found in a large number of ants, it appears that they play a major role in the social biology of the ants (Fig. 62A,B).

Fig. 62B. Leafcutter ants (*Atta* sp.) following a chemical trail across a piece of string back to the colony.

Ants have been shown to use chemicals in path finding. Some ants can be seen moving in files, one behind the other but not necessarily close behind. If the path is cleaned between two ants, the following ant stops when it reaches the cleaned part. Other ants following on the same path also stop and there may be some delay before any cross the cleaned area and continue on the path. Experiments show that ant trails are due to scent deposited on the ground from specialized glands. It is likely that ants use their antennae in detecting the scent and careful observation suggests that the trail exists not on the ground but in the air as a tunnel of scented air, a cross-section of which at right angles to the ant's path would be a semicircle (Blaney, 1977).

It is only in the ant of Myrmicinae that the glands associated with the sting have been utilized as the source of the trail pheromone. According to Wilson (1959) the contents of Dufour's gland of the fire ant, *Solenopsis saevissima* are evacuated through the sting in order to generate odour trails. On the other hand, several myrmicine genera lay odour trails with the poison gland which are used for defensive purpose (Blum *et al.,* 1964). In general, the trail pheromones appear to be relatively stable compounds which should be quite amenable to isolation and structural elucidation. Sheneirla and Brown (1950) have demonstrated that in nature *Eciton* trails retain their activities for several weeks. Similarly Blum *et al.* (1964) have shown that artificial trails prepared from attine poison glands retain their activities for several days at room temperatures. Beroza (1970) has demonstrated that n-hexane solutions of the trail pheromone are stable for at least 2 years when stored at 4°C. Moser and Silverstein (1967) have demonstrated that the trail pheromone of *Atta texana* was composed of at least two substances one of which was relatively non-volatile. The volatile fraction could not be detected by the ants after 60 min., whereas the non-volatile fraction was active for at least 6 days. It is presumed that the non-volatile fraction is oxidatively inactivated since artificial trails were highly active 4 months after preparation, provided that they were held under a vacuum of 0.5 mm of mercury.

Termite Trail Pheromones

In primitive as well as higher termites, odour trails are laid from a sternal gland in the ventral part of the fifth abdominal segment (Stuart, 1961). In the primitive genus *Zootermopsis* trails are laid to breaks in the nest structure and workers are thus recruited rapidly to a specific point in the nest. Alarm behaviour is also exhibited by *Zootermopsis* workers when they have discovered a nest break and the subsequent laying of an odour trail is part of a defensive reaction which results in building behaviour (Stuart, 1967). The number of workers recruited to the break is related to the intensity of the stimulus, the odour trail. In higher termites, food sources produce low-level alarm reactions when they are discovered and trails are laid from the food finds back to the nest. Thus, in both lower and higher termites, alarm stimuli release trail-laying behaviour (Beroza, 1970).

Bee Trail Pheromones

Lindauer and Kerr (1958) have demonstrated that the workers of many species of stingless bees, *Melipona* spp. (Meliponinae) lay trails from food finds to their nests. The trail is laid as a series of droplets which are placed by the scout bee at specific intervals on either vegetation or the ground as the bee moves from the food source to the nest. The distance between the scent marks varies with each species of bee but in all cases so far examined the marking pheromones originate in mandibular gland secretions (Lindauer and Kerr, 1960). Recruited bees orient to an aerial trail which results from the volatilization of the deposited mandibular gland droplets. These trails are rapidly reinforced by returning recruits and thus an effective concentration of trail pheromone is maintained during the critical period, the initial recruitment phase. The efficiency of recruiting in the higher Meliponinae is as great as that achieved by the honeybee *Apis mellifera*. In addition, stingless bee trails can be laid in the vertical component, whereas the honeybee is able to communicate only horizontal distances (Lindauer and Kerr, 1958). Therefore, stingless bees can recruit workers to flower concentrations in tall tropical trees, but honeybees must necessarily restrict their recruiting activities to the flora near the ground (Beroza, 1970).

ALARM PHEROMONES

Many species of social insects can rapidly communicate colonial disturbance by secreting pheromones from exocrine glands. Muschwitz (1964) has demonstrated that all major groups of hymenopterous insects employ alarm pheromones and Moore (1968) has shown that these excitatory substances are secreted by certain species of termites (Fig. 62C). The widespread distribution of alarm pheromones emphasizes the probable importance of these chemical releasers in the biology of social insects. Alarm pheromones appear to be least specific of the classes of chemical releasers of social behaviour.

Fig. 62C. Termite nasutes (*Nasutitermes* sp.) attracted by alarm pheromones to a hole in the nest.

Although alarm pheromones are far from species specific, it is nevertheless very clear that ants have a degree of olfactory acuity which enables them in the presence of their own excitatory odorant, to release alarm behaviour at lower threshold values than is possible with related compounds.

Alarm is provoked by the presence of a predator or in many social insects by a threat to the nest. The alarm behaviour of many eusocial insects is most familiar to humans when nest disturbance induces many ants, bees or wasps to an aggressive defence. Alarm pheromones attract aggressive workers and these recruits attack the cause of the disturbance by biting, stinging or firing repellent chemicals. Emission of more alarm pheromone mobilizes further defenders.

Alarm pheromones are also utilized as releaser of several different types of behaviour. Wilson (1958) has reported that the alarm pheromone of *Pogonomyrmex badius* functions as an attractant at low concentrations and a releaser of alarm behaviour at higher concentration. High concentrations of 2-heptanone release digging behaviour in excited workers of *Conomyrma pyramica*, a reaction also observed with workers of *P. badius*, which are exposed to their alarm pheromone. The alarm pheromone used by the stingless bee *Lestrimelitta limao* has been adapted to function as a key element in the robbing behaviour of this species. Workers of *L. limao* rob colonies of stingless bees in the genera *Trigona* and *Melipona* and their alarm pheromone "citral" (Blum, 1966) subserves several crucial functions during the robbing act. Citral which is released by the initial robbers, attracts other workers of *L. limao* to the nest which is under attack. Blum *et al.* (1969) have demonstrated that citral is highly effective in releasing escape behaviour in workers of stingless bee species normally raided by *L. limao*. A detailed account on this subject has been given by Beroza (1970).

THE QUEEN-SUBSTANCE

Among social insects the only sex attractant that has been identified is that of the queen honey bee *Apis mellifera*. The queen substance, 9-oxodecenoic acid is produced in the mandibular glands of the queen honey bee, which serves as a sex attractant to the drone during mating and keeps the workers in the hive informed of the continued presence of the queen. The secretion is obtained by beating the body of the queen and passed throughout the community by food exchange, so long as it is present the building of queen cell by the workers is inhibited.

Worker honey bees which are non-reproductive females do not produce queen substance in their mandibular glands. Butler and Simpson (1967) demonstrated that the keto-acid would attract swarming bees which had lost their queen. Further more, Butler *et al.* (1967) demonstrated that extracts of queens of *Apis cerana* and *Apis florea* were as attractive to drones as those of the *Apis mellifera* and further that the mandibular glands of *A. cerana* and *A. florea* produce the same chemical as the *A. mellifera*. It has been reported that in conjunction with another scent from the queen, the queen substance suppresses the maturation of the ovaries of the workers and yet another secretion keeps the workers in the swarm in close contact with queen (Wigglesworth, 1965).

Thus, it can be concluded that the "queen substance" produced by the mandibular glands of the queen bee and transmitted by food sharing to the worker bees, on which it has two effects (1) it prevents them from building the large queen-rearing cells and (2) it suppresses ovary development. This substance does not act alone; other factors, perhaps some volatile odour contribute to ovary inhibition and removal of the mandibular glands reduces the inhibition of queen-cell building but does not eliminate it completely. The active material is 9-oxodecenoic acid of which 0.13 µg per bee will inhibit queen-cell production. It also serves as a sex attractant in mating. But the mandibular glands contain other substances; one of these, probably 9-hydroxy-decenoic acid has been recognized as the pheromone, which prevents the cluster of the bee swarm from breaking up (Wigglesworth, 1965).

TERRITORIAL PHEROMONES

It has been pointed out that the male bumble bees *Bombus* spp. use their mandibular gland secretions in order to mark territorial sites which may play an important role in their mating biology. While the scent marks produced by a male of one species do not attract males of other species and therefore function as territorial markers. The marks do attract both males and females of the same species thus bringing the reproductives of both the sexes together and increasing the probability that mating will occur. Probably the scent marks also physiologically initiate the males and females for subsequent copulatory activity and thus play the role of sexual stimulants (Beroza, 1970).

Bergstrom *et al.* (1968) have identified 2, 3 dihydro-6-trans farnesol as the main component in the mandibular gland secretion of *Bombus terrestris*. Several other chemicals have been isolated from the mandibular glands of different species of *Bombus* (see : Beroza, 1970).

AGGREGATION PHEROMONES

The release of an aggregation pheromone causes conspecific insects of both the sexes to crowd around the source of the pheromone. Aggregation may lead to increased likelihood of mating but in contrast to many sex pheromones both sexes may produce and respond to aggregation pheromones. The potential benefits provided by the response include security from predators, maximum utilization of a scarce food resource, overcoming of host resistance and attraction of social insects as well as to the chance to mate.

SPACING PHEROMONES

Gillott (1995) remarked that a relatively new discovery in pheromone research are those pheromones that stimulate insects to spread out so as to maintain an optimal population density. They may be produced by immature and adult insects and may be olfactory or tactile. Spacing pheromones appear to have been best studied in relation to oviposition. For example, after laying, the female apple maggot fly, *Rhagoletis pomonella* releases, a pheromone from her ovipositor that deters oviposition on the fruit by other females. Several Lepidoptera also produce oviposition - deterring pheromones which limit the number of eggs laid on a given plant. Further, the presence of feeding larvae of *Pieris brassicae* inhibits egg laying, suggesting that the larvae also produce a pheromone. Thus, the pheromones which control density of the insect population are called spacing pheromones or dispersion pheromones. Spacer pheromones of colony-specific odours may be used to ensure an even dispersal of colonies of conspecifics, as in African weaver ants, *Oecophylla longinoda* (Formicidae : Hymenoptera) (Gullan and Cranston, 1994).

CASTE REGULATING PHEROMONES

The regulation of caste differentiation in social insects is a morphogenetic phenomenon, just as are the changes from larva to larva, larva to pupa and pupa to adult. Such changes depend on the activity of corpora allata (level of JH in the haemolymph) for their manifestation. It is not surprising therefore, to learn that the pheromones that regulate caste differention including development of reproductives, exert their effect ultimately via corpora allata. In lower termites, e.g. soldier formation can be induced in experimental colonies by administration of juvenile hormone through feeding, topical application or as vapour. Thus, the soldier-inhibiting pheromone may act by inhibiting the corpora allata or by competing with juvenile hormone at its site of action (Luscher, 1972).

SEMIOCHEMICALS

Gullan and Cranston (1994) have remarked that the communication chemicals (semiochemicals) may function between individuals of the same species in which case they are called pheromones, or between different species in which case they are called `allelochemicals'. Interspecific semiochemicals may be grouped according to the benefits they provide to the producer and receiver like kairomones, allomones and synomones.

Kairomones

These are the communication chemicals that benefit the receiver and disadvantageous to the producer, for example myrcene, the terpene produced by a ponderosa pine when it is damaged by the western pine beetle, acts as a synergist with aggregation pheromones, to lure more beetles. Thus, myrcene and other terpenes produced by damaged conifer trees can be kairomones (i.e. the chemicals that disadvantage the producer), in this case by luring damaging timber beetles. A kairomone need not be a product of insect attack e.g. elm bark beetles - *Scolytus* spp. (Scolytidae) respond to α - cubebene, a product of the Dutch elm disease fungus, *Ceratocystis ulmi* that indicates a weakened or dead elm tree. As remarked by Gullan and Cranston (1994) the elm beetles will inoculate previously healthy elm trees with the fungus but pheromone - induced beetle aggregation form only when the kairomone (i.e. fungal α-cubebene) indicates suitability for colonization. Host plant detection by phytophagous insects also involves reception of plant chemicals that therefore act as kairomones.

Allomones

These are the chemicals that benefit the producer but have neutral effects on the recipient e.g. defensive or repellent chemicals are allomones that advertize distastefulness and protect the producer from lethal experiment by prosepctive predators. The effect on a potential predator is considered to be neutral since it is warned from wasting energy in seeking a distasteful meal.

Synomones

As stated, the terpenes produced by damaged pines are kairomones for pest beetles but if identical chemicals are used by beneficial parasitoids to locate and attack the bark beetles, the terpenes are acting as synomones i.e. benefiting both the producer and the receiver (Gullan and Cranston, 1994). Thus, according to the above authors, the α-pinene and myrcene produced by damaged pines are kairomones for species of *Dendroctonus* but synomones for the pteromalid hymenopterans that parasitize these timber beetles. In a similar manner α-cubebene produced by Dutch elm fungus is a synomone for the braconid-hymenopteran parasitoids of elm bark beetles (for which it is a kairomone) (Gullan and Cranston, 1994). It has been further remarked by Gullan and Cranston (1994) that if a damaged host plant produced a repellent odour then the chemical would act as :

1. An allomone that deters non-specialist phytophages.
2. A kairomone that attracts a specialist phytophage.
3. A synomone that lures the parasitoid of the phytophage.

DEFENSIVE SECRETIONS

Furthermore, many insects produce "defensive secretions" which are repellent to potential predators. These comprise a vast range of saturated and unsaturated hydrocarbons and their derived alcohols and aldehydes (ranging from hexenal and tridecene to pentadecane), fatty acids such as butyric and caproic acids, a great number of phenolic substances and quinones such as phenol, *m*-cresol, benzoquinone, toluquinone etc. and a variety of terpenoids such as geraniol, farnesol, citral, citronellal etc. These different substances perform different functions in different insect species.

Wigglesworth (1970) has summarized the work on pheromonal control of aggregation, dispersal, trail formation and social coherence. According to Wigglesworth (1970) defensive secretions as such would not fall within the definition of pheromones. But what are commonly regarded as repellent or defensive secretions against outsiders, can be utilized within the species for recognition, aggregation and trail formation. The trail pheromone of *Zootermopsis* contains many compounds, but one of the most active is caproic acid. The "Nassanoff glands" of the honey bee worker secrete a mixture of terpenoids (viz. citral, geraniol etc.) and this odour is used by foragers as an attractant to secure recruits for their expeditions to rich collecting grounds. In *Lasius niger* and Ponerine ants generally, the contents of the stomach and hind gut will serve as an effective scent trail; in Myrmecine ants the secretion of the poison gland is used.

Defensive secretions may also play the opposite role. They may serve to alert and disperse aggregations. That is seen in *Formica rufa* where formic acid is a potent means of defence and offence but serves also to alert the colony to a threatened attack and so to induce a general agitation throughout the nest. The citral produced by the mandibular glands of the leaf cutting ant *Atta* likewise, serves this same dual purpose of repellency and alarm and in the cotton stainer *Dysdercus,* which produces a mixture of hydrocarbons ranging from hexenal to pentadecane, this product can serve both as a repellent against predators and as a warning signal that causes dispersal of the aggregations which occur in this species.

A very large number of alerting pheromones have been isolated and identified in ants and termites, many are terpenoids, others are aliphatic ketones and esters. In the fire ant, *Solenopsis*, a whole series of complex acts of behaviour, mass foraging, trail following, colony emigration, alarm, grooming and clustering all are largely induced by chemical (i.e. pheromone) releasers of this kind. The mandibular glands of the queen honeybee contain a variety of organic acids, one of which is 9-hydroxy-decenoic acid, which serves as ``cohesion pheromone", which keeps the swarm of worker bees clustered around their queen. There is indeed evidence for the presence of widespread pheromones in other social insects notably ants and termites (vide supra) which are distributed throughout the colony by food sharing between adults and serve the purpose of colony cohesion but the chemical nature of these materials is not well known.

Among the social insects one of the best known chemical messengers (i.e. pheromones) of this kind is the queen substance, 9-oxodecenoic acid also secreted by the mandibular glands of the queen bee. This secretion is smeared over the integument of the queen during her grooming activities. The queen is usually surrounded by a circle of workers, all facing her and licking her and passing on the material in the course of food exchange with other workers. This is probably the chief means by which workers are kept informed of the presence of the queen. If the queen is removed, a change in behaviour among the workers becomes apparent and within 24 hours they begin building large "queen cells" to rear new queens. At the same time the ovaries of many of the workers ripen and they begin to lay eggs.

It may be that the absence of the same queen substance, in combination with a special queen scent is responsible for both these effects. The queen scent is a product of glandular cells in the mid dorsal area of the abdomen. There is no evidence that queen substance or odour act directly on the reproductive system, it seems more probable that they act upon the sense organs and the central nervous system and thus cause indirectly a change in hormone secretion. During the first few days after the removal of the queen the corpora allata of the worker bees increase in volume, their activity seems normally to be inhibited by the queen substance (see: Wigglesworth, 1970).

Further, I would like to mention that Blomquist and Vogt (2003) have described the biochemistry and molecular biology of insect pheromones in considerable detail and Conte and Hefetz (2008) have given a comprehensive account of the role of pheromones in honey bees and a variety of ants. Readers interested in the subject must consult the above work.

16

GENETIC CONTROL OF BEHAVIOUR AND EVOLUTION OF BEHAVIOUR

The discipline of behaviour genetics is quite young and credit for establishing behavioural genetics goes to John Fuller and W.R. Thompson with the publication of their book `Behavioural genetics' in 1960. In the past few decades research on genetic influence on animal behaviour has progressed at a rapid speed in various lines. The survey of literature on evolution of insect behaviour has clearly revealed two important features viz. (1) diversity of behaviour and (2) adaptiveness of behaviour in various circumstances which has been briefly described in different chapters.

A. GENETIC CONTROL OF INSECT BEHAVIOUR

There are many example of the demonstration of genetic control of insect behaviour. A great deal of work has been done with *Drosophila melanogaster* and other *Drosophila* species as well as a number of other kinds of insects. An extensive treatment of the genetics of behaviour is far beyond the scope of this text. The readers are referred to McClearn and DeFries (1973). Ehrman and Parsons (1976) for a comprehensive information.

There are several examples of gene mutations in *Drosophila* that change a behaviour pattern. An early example was a sex linked mutant with yellow body colour instead of the normal gray. Males of these yellow mutants were less successful in mating than the gray wild type. In rare instances, it has been possible via breeding experiments to account for particular behaviour patterns on the basis of the action of single gene. An outstanding example of this is the work of Rothenbuhler (1964). He studied two inbred lines of honey bees one that had been selected for resistance to American `foulbrood' - (a disease of honey bee larvae) and another that was susceptible. He noted a distinct behavioural difference between the two lines, the resistant line removed foulbrood - killed larvae relatively soon after death (i.e. hygienic behaviour) and the susceptible line did not (non-hygienic behaviour). Genetic analysis revealed that the action of two sets of dominant/recessive genes on different chromosomes would account for the results of breeding experiments, between the two lines. One set of alleles involved uncapping of the cells in which diseased larvae were developing : recessive alleles, uncapping and dominant alleles, no uncapping. Thus the homozygous recessive displayed uncapping behaviour. Homozygous dominants and heterozygous did not. The other set of alleles involved removal of diseased larvae : recessive allele, larvae removed and dominant alleles, diseased larvae not removed. Homozygous dominants and heterozygotes did not remove diseased larvae.

It appears that most behaviour patterns are probably under polygenic control and *directional selection experiments* have been useful in establishing whether a gene behaviour has a genetic basis (Ehrman and Parsons, 1976). It has been remarked by Ehrman and Parsons (1976) that in selection experiments individuals are selected at high or low extremes of a distribution in the hope of forming separate high or low lines in subsequent generations. If a trait does not have a genetic basis, then attempts to select for extremes would fail. A good example of a selection experiment is the work of Hirsch and Erlenmeyer-Kimling (1961). They showed that with selection of *Drosophila melanogaster* that had previously been assumed to be negatively geotactic could be separated into both positively and negatively geotactic lines. Following establishment of extreme strains, it has been possible in some cases to evaluate the degree of involvement of different chromosomes e.g. in the case of geotaxis in *Drosophila* genes on each of the three major chromosomes were found to exert an influence (Hirsch and Erlenmeyer-Kimling, 1962). In *Drosophila,* several other behavioural traits have been shown to be under polygenic control, including locomotor activity, chemotaxis, duration of copulation, optomotor response, phototaxis and so on.

Further, Hoy and Paul (1972) and Hoy (1974) have found that the sound produced by adult male crickets are not learned but genetically controlled. Furthermore, hybrids produced by crossing *Teleogryllus* and *Gryllus* species exhibit song pattern that are intermediate between those of the parents. *Teleogryllus* songs are controlled by several different genes on different chromosomes.

Thus, it may be pointed out here that the behaviour genetics can play an important role in the future development of the science of insect behaviour. There are three ways in which this can occur. The first is through the impact of behaviour genetic- research itself. There are many exciting phenomena whose genetic bases are under study and whose impct will be added to the field of insect behaviour as a whole. The second is in providing valuable experimental preparations. The behaviour genetics can provide tools such as genetic mosaics, strains with different brain weights, and techniques for analysing developmental effects that can be of great value to animal behaviourists. Finally, behaviour genetic research provides important cuations regarding the kinds of orderliness we can expect in formulating the laws of behaviour.

Strain Differences

The basic strategy for using strain differences to determine whether genes affect behaviour is simple. There are two major factors in the development of behaviour viz. genes and environment. The strategy to locate organisms of different genotypes hold environmental factors constant and determine whether the genotypic factor influence the behaviour.

Genes and Biochemistry

Biochemically, the gene may be regarded as a portion of a DNA molecule that codes for the formation of a particular polypeptide. DNA molecules of the cell nucleus convey specificity to RNA. The RNA directs biochemical activity in the cell's cytoplasm that ultimately results in the formation of structural proteins and functional enzymes (see : Dewsbury, 1978). The enzymes are catalysts for biochemical reactions whose rates of activity they alter. These chemical reactions affect the skeletal systems, muscles, endocrine glands, the peripheral and central nervous systems and indeed all systems of the body. It is clear that there must be a precise set of pathways leading from genes to enzymes to brain chemistry and behaviour.

Much of the work on the genetic control of animal behaviour particularly in mammals have been conducted in the past few decades but only a little is done on the insects, however, there is no basic difference in principles guiding the genetic control of behaviour in insects.

B. EVOLUTION OF BEHAVIOUR IN INSECTS

The tremendous diversity of behaviour of different species not only among distantly related species but also among closely related species as well, has been very clear, if we study the insects of each species which have become adapted to live a particular `life style' is appropriate for a given environment i.e. they have become adapted. The most parsimonious explanation for the diversity and adaptedness of behaviour and indeed of all biological phenomena comes from the theory of evolution.

Dewsbury (1978) has discussed the evolution of behaviour of animals in general. Here, the information collected from the work of Dewsbury and other sources on evolution of behaviour with special reference to insects, is briefly described.

Evolution of Behaviour at Species Level

Statements about behavioural evolution are most precise when related to a clearly defined and measurable behaviour pattern in the restricted group of species i.e. at the species level. Lineages are most clear when a graded sequence of species is available for study so that probable graded steps in the evolution of the behaviour can be inferred. Inferences based on groups of species where such graded steps are available have been termed `strong' comparative inference, as opposed to inferences where such graded sequences are lacking which may be termed "weak" comparative inferences. The graded sequences of evolutionary trends seen among closely related species have been termed *ethoclines.*

Lindauer (1961) used the species level of comparative analysis to study the evolution of the complx dances that signal the location of food sources in honeybees. A plausible sequence was recreated from the stingless bees, in which the guidance of a leader and odour markers are necessary for the transmission of information, through bees of different species of the genus *Apis* in which the direction is coded first directly with respect to the sun and later indirectly relative to gravity.

Evolution of Behaviour at Phyletic Level

It has been pointed out by Dewsbury (1978) that the comparisons at the phyletic level are often more difficult than those at the species level, because the species under study differ in so many ways that : (1) it is often impossible to be sure that they are studied under comparable conditions and (2) once differences are observed, it is difficult to know which of the many factors that differentiate the species are associated with the particular difference of interest. The major advantage of comparisons at the phyletic level is that they allow one to reconstruct the broad sweep of evolutionary change occurring in a particular group of species, Thus, while comparisons at the phyletic level may be of little value for elucidating the nature of evolutionary processes, they can be of great interest in constructing the broad evolutionary history of behaviour.

There are many examples of the application of comparisons at the phyletic level. Alexander (1964) traced the history of evolution of mating behaviour in Arthropods and found it related to the evolution of spermatophores, adaptation of aquatic or terrestrial habitats and more specific environmental conditions within particular groups.

The Micro Evolution of Behaviour

The study of micro evolution is concerned with small amounts of evolutionary changes as for example in the earlier stages of differentiation of behavioural patterns. Some different kinds of quantitative changes among species can be observed in the micro evolutionary study.

Changes in Threshold for the Elicitation of Homologous Behavioural Patterns

Micro evolutionary changes can produce differences in the threshold for elicitation of particulr homologous behavioural patterns e.g. moths of different species differ with respect to the intensity of stimulus required to elicit defensive displays. The more vividly coloured, unpalatable species respond to the slightest touch with movement and a display. By contrast, the more cryptically coloured species remain motionless and only display in response to a hard or sharp prick (Blest, 1957).

Changes in the Frequency with which Homologous Behavioural Patterns are Performed

With changes in frequency there is a common repertoire of behavioural patterns between closely related species, but the species differ in the relative frequency with which they display the behavioural pattern. For example, two closely related species of fruit flies *Drosophila melanogaster* and *Drosophila simulans,* each show two patterns of wing display during courtship (1) scissoring (in which both wings are slowly opened) and (2) vibration (in which one wing is extended and rapidly vibrated in the vertical plane). Vibration is the dominant display among *Drosophila melanogaster* whereas scissoring is more common in *Drosophila simulans.* Indeed, the casual observer may never observe scissoring in *Drosophila melanogaster,* it revealed only with prolonged observations (Manning, 1959; Dewsbury, 1978).

Changes in the Speed with which Homologous Behavioural Patterns are Performed

Differences in the speed of homologous behavioural patterns imply that while the form of the movement is unchanged, the movement occurs at a faster or slower pace e.g. the protective displays of moths (Blest, 1957).

These quantitative changes in behaviour occurring at the micro evolutionary level are summed over time and generations to produce the qualitative differences that so characterize evolutionary changes across species (Dewsbury, 1978).

Reproductive Isolation

Flies (*Drosophila*) of two closely related species or of two similar genotypes differing in some prominent marker gene are intermixed and permitted to mate. Such flies can either mate with partners of their own genotype or form hybrid matings. The offspring of hybrid matings are removed and not permitted to reproduce. If this procedure is followed over several generations, reproductive isolation between the populations can be seen to develop, whereas a relatively large proportion of matings in the first few generations and in control lines may be hybrid matings. The percentage of matings in the selected lines that produce hybrids decreases over 20 to 40 generations (Dewsbury, 1978).

Crossley (1974) analysed the details of the reproductive interactions between her ebony (usually black) and vestigial flies after selection against hybrids. She found that courtship interactions of her two populations had been changed over the course of her experiment. There occurred (1) an increase in the tendency of females to resist courtship from males of the strange (i.e. heterogametic) strain and (2) a greater tendency of males to respond to the repelling movements from females of the foreign strain by terminating courtship than to respond similarly to the repelling movements of females of their own strain.

Selective Pressures on Displays

Cullen (1966) has remarked that a number of selective pressures act on behavioural patterns that serve as signals and these pressures are responsible for the reinforcement and change of the signals through evolution. Thus, evolutionary changes act to make signals more conspicuous and decrease ambiguity.

According to Cullen (1966) there are three fundamental classes of selective pressures viz.

(1) *Between species* - Reduced ambiguity of signals functions to encourage reproductive isolation and prevent hybridization. It also acts to reduce aggressive encounters in closely related species not competing for identical ecological resources.

(2) *Within species* - Within species, it is advantageous for signals to be as clear as possible so as to minimize confusion among different signals.

(3) *Individual differences* - Signals can serve to identify the individual emitting the display.

Evolution of Courtship in Balloon Flies

The empid fly, *Hilara sartor,* displays an interesting courtship pattern. Males construct a silk `balloon' about as large as they are. They then join other males in a swarm, from which a female selects a mate. The female accepts her mate's balloon and the pair depart to mate (Kessel, 1955).

Kessel (1955) studied a number of species from among the thousands of the family of empid flies and found a graded 8 stage sequence in different species corresponding with the probable evolution of this unusual courtship pattern viz. :-

1. Males search for a female and court her in isolation from other pairs.
2. Males capture a prey, present it to a female and she consumes it before mating.
3. Males capture prey from groups and present females who receive the prey and mate.
4. Males mate as in stage 3 except that a few strands of silk are applied to the prey prior to swarming.

5. Males mate as in stage 4 except that males wrap a complete silk covering about the prey.
6. Mating is the same as in stage 5 except that males consume the edible portion of the prey before presenting it so that the female receives a non-nutritious husk.
7. These species are not carnivorous, unlike those of the previous 6 stages. A dried insect fragment is collected and used as the base for the weaving of a balloon.
8. Males omit the insect fragment in the formation of a balloon, as in the case of *Hilara sartor*.

The process of displays of progressive changes among closely related forms are readily apparent in the above example.

17

COMMUNICATION IN INSECTS

This topic is of vital importance in insect ethology. A variety of behavioural patterns are influenced by the communication systems hence in many chapters, a brief account of the subject may be found.

Communication between insects involves action on the part of one individual which alters the behaviour of another individual in a predictable way. This rather broad definition encompasses many modes of communication - visual signals, touching, stroking, tapping, grasping, the production of flashes of light, pulses of sound, puffs and squirts of chemical - which evoke responses ranging from simple recognition to the sophisticated coordination of activity which we have seen in the social insects. Within the social insects, this repertoire of communicative behaviour may be extended to include elaborate and sometimes bizarre effects such as the exchange of chemical signals conveyed in liquid food which results in the control of caste development and even in programmed execution and cannibalism.

The signals produced by an insect may be of significance to, and may be recognized by, many different species - that is, they are extraspecific. Usually, such signals are concerned with defence and warning, perhaps frightening off an attacker or giving a general warning of the presence of a predator. Conversely, intraspecific signals have significance only for a single species and are largely ignored by members of other species. They are thus commonly used in courtship where they reduce the possibility of interspecific mating and set up psychological barriers keeping the species apart. In addition to their use in courtship, intraspecific signals may also be involved in sexual isolation, aggregation, aggression and, in social insects, behaviour such as foraging, recruitment and recognition of nest-mates (Blaney, 1977).

In communication, the information conveyed by the signals is in the form of a code of stimuli using one or more sensory modes which can be produced by the sender and received by the receiver.

In this chapter, a detailed account of the communication system as seen in insects is given. The communication in insects takes place in many ways either singly or in combination of visual, auditory (acoustic), tactile or chemicals.

1. VISUAL COMMUNICATION

As remarked somewhere else, the most interesting visual communication system is that of the photogenic insects, specially the fireflies or glow worms. These produce light flashes that are species specific and meant for mating. It has been found that timing and pattern of flashes bring the sexes together. In summer/rainy season we are very well aware of flying fire flies males, each flashing species has a specific code to which females respond. If the female fire fly responds in the appropriate code, the fly moves towards its source (Lloyd, 1971). Lloyd, J.E. (1965) recounts a remarkable case in which females of *Photuris* a predacious species lure smaller males of *Photinus* to dinner by mimicking the flashes of the latter's code. Moreover, female *Photuris* mimic at least four smaller fireflies and thus can vary their menu. The arboreal males of fireflies, flash synchronously lighting up whole trees which appear extremely beautiful like christmas trees. These are apparently mating aggregations. The visual signals are involved in reproductive behaviour, aggressive displays, and in defensive alarm displays etc. (Lloyd, 1971, 1977).

Light production in insects usually has a sexual significance. The male fire-fly *Photinus pyralis* (Fig. 62D) exposes light-producing cells on his body and emits flashes of light at intervals of about 5-8 seconds while flying some 50 cm above the ground. Females perch on top of low vegetation and if a flashing male comes within about 2m the female flashes in response. The female flash is emitted approximately 2 seconds after that of the male and the exact delay between flashes is characteristic of the species. The male approaches any light flash which shows the right time interval, ignoring all others, and as a result of further flashing is led right up to the female. The time relationship between transmission and answer is the all-important characteristic of this signal. Certain species of fire-flies in Burma and Siam form groups which flash synchronously, but the control of this remarkable behaviour is not understood.

Fig. 62D. Ventral view of firefly (*Photinus pyralis*) showing abdominal luminescent organ

In the following pages, mechanism of the production of light in insects is given. The light emitting mechanisms studied in the lampyrid firefly, *Photinus pyralis* may be typical of luminescent Coleoptera. The enzyme luciferase oxidises a substance, luciferin in the presence of an energy source of adenosine triphosphate (ATP) and oxygen to produce oxyluciferin, carbon dioxide and light. Variation in ATP release controls the rate of flashing and differences in pH may allow variation in the frequency (colour) of light emitted. The luminescent organs may be located in different bioluminescent insects, in different parts of the body and the light emitted may be white, yellow, red or green.

Further, it may be clearly remarked that luminescent as with all communication systems, provides scope for abuse, in this case, the luminescent signal of courtship between beetles is misappropriated. Carnivorous female lampyrids of some *Photuris* species is an example of aggressive foraging mimicry, imitate the flashing signals of females of up to five other fireflies species. The males of these different species flash their responses and are deluded into landing close to the mimetic female upon which she devours them. The mimicking *Photuris* female will eat the males of her own species but cannibalism is avoided or reduced since the *Photuris* female is most piratic after mating at which time she becomes relatively unresponsive to the signals of males of her own species (Gullan and Cranston, 1994).

It may be pointed out that ants using visual navigation usually orientate to distant landmarks, often the sun, which they sense with their eyes. This was strikingly demonstrated by an experiment of Santschi's carried out under natural conditions. An ant running in sunshine was shaded by a small screen and at the same time the sun's light was reflected on to it from the opposite side by a mirror. The ant turned and ran in a new direction so that its body lay at the same angle to the reflected light as it had previously lain to the sun.

These menotaxes or light-compass reactions, in which the insect's body is maintained at an angle to the light, are a development of the simple phototaxes in which the insect's body points directly at or away from the light. Inexperienced ants which are just learning to forage use phototaxes, but later they become able to use menotaxes and also, of course, to compensate for the fact that the sun's position alters in the sky by about 15° in each hour. Experienced ants can also follow a crooked course outwards and return by a course which is more or less the shortest straight line (see Blaney, 1977).

Thus, among visual communication systems exists a great variety of patterns and colourations which when combined with appropriate behavioural display elicit a response. Kaiser (1974) demonstrated that a visual display stimulates the chasing response. Kaiser also built a model dragonflies and found that dragon flies respond most quickly to patterns of their own species. Many insects use visual communication systems combined with tactile and chemical communication in location and recognition of mates.

2. AUDITORY OR ACOUSTIC COMMUNICATION

Sounds are produced by many insects in the course of their day-to-day activities and may have little or no signficance in the life of the insect. However, many sounds do have significance as signals. These sounds are produced by special mechanisms and are detected by tympanal organs - the equivalent in insects of our ears. In some Orthoptera and Coleoptera, sounds are commonly produced by frictional mechanisms. A ridge is scraped over a series of grooves on another part of the body which is thus caused to vibrate. In some Homoptera and Lepidoptera, on the other hand, a thin membrane is set vibrating by the direct action of special muscles (Blaney, 1977).

Sound signals which are used in intraspecific communication usually consist of discrete bursts of sound separated by intervals of silence and repeated in a regular manner. Such are the songs of insects, and some species have a variety of different songs for different occasions. The songs differ in the timing of sounds and are used most commonly in courtship but some songs also lead to aggregation of particular species and separation of different species.

Most grasshoppers and locusts (Orthoptera) produce sounds by rubbing a ridge on the inside of the hind leg against a knobbly vein on the outside of the folded wing. This friction causes the wing to vibrate at its own resonant frequency which varies from 2-50 thousand cycles per second. Each stroke of the leg produce a single pulse of sound.

Often female grasshoppers are capable of singing too but do so only when they are sexually mature. The courtship behaviour of the British grasshopper *Chorthippus brunneus* is fairly typical. The male sings his calling song and if a responsible female hears it she sings a similar song in reply. By pausing to listen to each other's sounds, the two insects are able to orientate to each other and move closer. When the male sees the famle he starts to sing his courtship song. If, however, another courting male intervenes, the two rivals sing an aggressive song at each other until one of them gives way and withdraws. Whilst singing the courtship song, the male hops towards the female, eventually he hops onto her and attempts to copulate. If he is rejected he starts the courtship song again. If, during copulation, the female becomes restive or tries to move away, the male sings a copulation song which calms her and allows copulation to progress (see Blaney, 1977).

It is a matter of common experience that insects of several orders produce sounds that attract individuals of the same species. The male mosquitoes move towards the sound of the female's buzzing wings. Several Diptera and Hymenoptera are attracted by the sound of the wings of the opposite sex. Many beetles produce sound by stridulation. These sounds are species specific and serve for sexual recognition e.g. Cicadidae, (Homoptera), and Acrididae (Orthoptera). In both the groups it is predominantly males that ``sing'' and these sounds are long distance attractants for females. Females grasshoppers and katydids rarely make mistakes in locating mates of the appropriate species.

A different method of sound production by friction is used by crickets and long-horned gasshoppers. In these insects the fore wings or elytra are somewhat hardened and parchment-like in consistency and act as shields for the more membranous hind wings. Each elytron has a vein on the underside modified to form a toothed file (Fig. 62E) while on the edge of the opposite elytron there is a ridge which acts as a scrper. In sound production, the elytra are raised above the body at an angle of about 45° and opened and closed like the blades of a pair of scissors. This causes one of the scrappers to rasp along one of the files causing the elytron to vibrate and produce a sound. The sound thus produced can be very penetrating; and can be heard over a distance of 1.6 km. It is interesting to note that the sounds are often quite musical and in some parts of the world crickets (Fig. 62 F) are sold in cages to be kept as singing pets (Blaney, 1977).

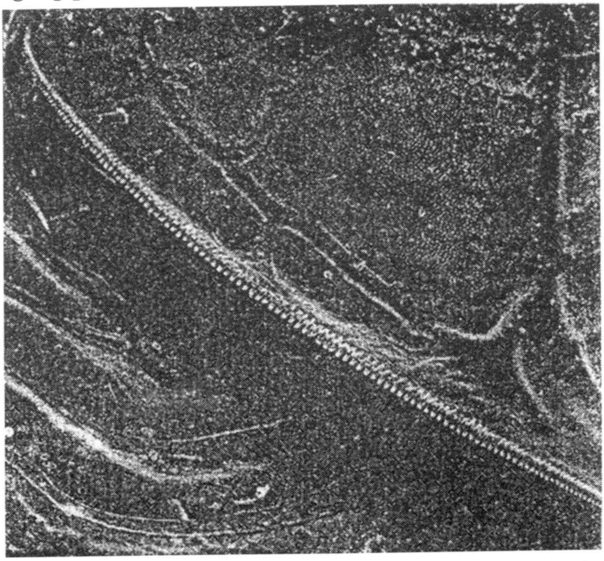

Fig. 62E. File on front wing of a male cricket (*Acheta assimilis*) that is rubbed across a scrper on the other forewing as the wings are moved. The result is vibration of the wings and a sound that attracts female cricket.

In Diptera e.g. mosquitoes, wing vibrations by female produce the sound and male perceive these through antennal receptors. The antennae bear dense whorls of hair (Fig. 62G) that are sensitive to vibrations in the air and they detect the sound made by the wing beats of the females.

Several species of Orthoptera and in Cicadidae, singing males aggregate and produce synchronous `choruses' capable of attracting females over a much longer distance than could the song of an individual. Apparently a leader sets the initial pace and others nearby respond when he calls. The resulting chorus ranges in size from antiphonally stridulating pairs in male katydids to groups of several thousand in periodical cicadas. Aggregation probably increases the frequency of mating by drawing more females. It has been found that singing Orthoptera and Cicadas produce different types of sounds and not all of which are sexual. There are also alarm sounds that cause receiver to cease singing and to move away from the alarm caller.

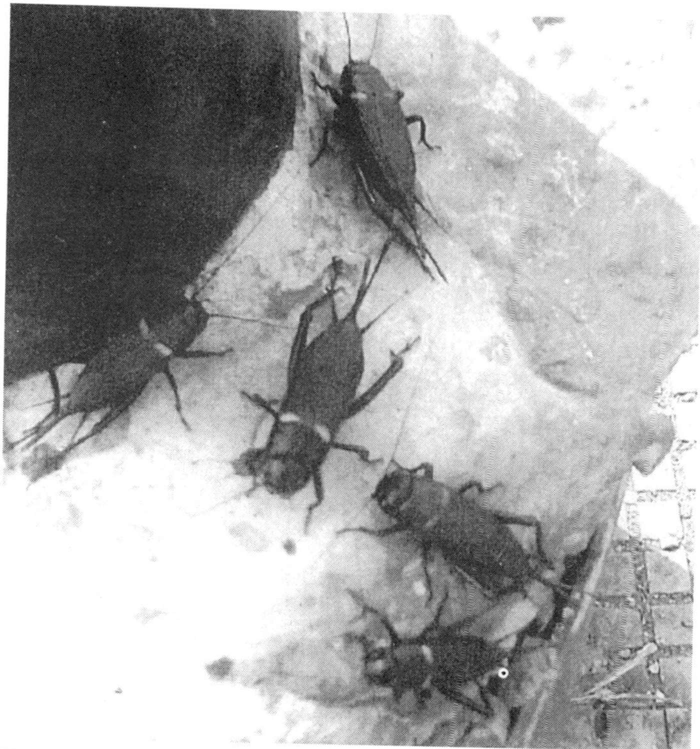

Fig. 62F. Part of a large group of crickets. These crickets produce a loud and musical sound.

The members of the order Orthoptera, Hemiptera (both Heteroptera and Homoptera), (3) Lepidoptera, (4) Coleoptera and (5) Diptera are known to hear and to produce distinctive sounds. Interspecific sound communication tends to be mediated by unpatterned sounds, such as ``warning" sounds while intraspecific communication typically involves highly patterned sounds. As intraspecific "signals", sounds are involved in aggregation, sexual behaviour, aggression, alarm and social interactions. Acoustical signals are very effective in the sense that they can operate over relatively long distance, can be modulated in many difficult ways, can operate in dark, can provide directional information etc.

Fig. 62G. A male mosquito antenna emerging from cup-like bases showing whorls of sound sensitive hair.

It has been pointed out by Horn (1978) that control of singing is genetic in that each song is species specific. Many sound producing insects have a circadian rhythm of sound production, commencing at sundown (crickets and some grasshoppers) or singing only in day light hours e.g. Cicadas.

3. TACTILE COMMUNICATION

In this case the tactile sense organs are stimulated and a response is elicited by them. It occurs in all insects at least at mating time together with chemical and visual communication is a principal link in sexual recognition. Social insects often strike and groom one another with antennae and mouth parts and termite soldiers alarm a colony to the presence of intruders by unison tapping of mandibles against the wall of the termitarium producing a sound of the same pitch. Termites throughout the termitarium (nest) perceive these taps tactilely through their tarsi.

Perhaphs the most involved tactile communication system is the wagging dance of the honey bee. A ``scout" bee that has located a new food source, returns to the hive and performs a figure eight (8) dance on the vertical comb inside (see : Fig. 29B), during which she waggles her abdomen while walking upward. Because the bees are crowded together on the comb, near by bees can feel the direction of the dance. Frisch (1974) demonstrated that the departure from vertical of the angle of the dance is equal to the angle of departure of the food source from the sun, and that the length of time spent per wagging (waggle) walk is proportional to the distance of the food source. The most remarkable fact is that the bee may do a Marathon dance for over three hours on the comb (Frisch, 1967), during which she compensates for the apparent movement of the sun through the sky by changing the waggle angle on the comb. According to Lindauer (1965) this adjustment is done in the total darkness of the hive without a peep at the sky, it is probably, a function of the bee's ``internal clock" and its conditioning to the movement of the sun. Hive bees, perceiving duration and direction of the dance, leave the hive at the appropriate angle and fly the correct distance directly to the food source.

4. CHEMICAL COMMUNICATION

Chemical communications by far the most common form of communication among insects. A given chemical signal may be a single compound or a mixture of compounds and may have been synthesized entirely in the insect or may have been obtained from the environment specially from food.

Chemical signals can be classified on the basis of whether they mediate intra- or inter specific communication. Chemicals that function intra-specifically are called pheromones. Much of the insect behaviour is mediated by pheromones, the chemicals secreted by other insects of the same species (see : Chapter 15). Pheromones are secreted into the environment by exocrine glands or by some other structures e.g. scent scales or wings (Lepidoptera) and are percieved by chemoreceptors present on antennae, tarsi or mouth parts of a recipient. Such chemical messengers usually volatilze or disperse rapidly in the atmosphere, serving as either releasers for an immediate behavioural response like the sex pheromones of female moths that induce males to fly upwind or as primers for longer lasting physiological changes in recipients like the pheromones that determine caste in social insects. Pheromones are used in extremely small quantities, stimulate activity over amazingly long distances.

A remarkable example of the power of pheromones to elicit a response at a distance and when in very low concentration is found in the sex attractants in butterflies and moths (Lepidoptera). The famous French naturalist Fabre described how in many species of moths, males would assemble round a female or even round an empty box which had contained her. His claim that the males came from miles away, attracted to her scent, was hard to believe when the small amount of chemical a female could produce was considered in relation to the large volume of air into which it would have to be released. Experiments have subsequently shown, however, that Fabre was right. The males come from downwind and the presence of pheromone stimulates upwind flight so that they are guided to the female by the wind and not by a concentration gradient of the very feeble chemical stimulus. Only when they are very near to the female does the concentration of chemical become sufficiently high for the male to follow the concentration gradient. A concentration of pheromone of only 3,250 molecules per cubic inch (200 per cu cm) is adequate to stimulate upwind flight in the male of the moth *Bombyx mori*. The female stores about 10^{14} molecules of the attractive substance called bombycol in her gland reservoirs and this is enough to produce a threshold concentration in a volume of air of elliptical shape, drawn out several miles downwind (Blaney, 1977).

Thus, aggregation pheromones draw more than two individuals together. Virgin bark beetle females (Scolytidae) release pheromones that attract both sexes to mating and oviposition sites thus aggregating large number of beetles to recently damaged timber (Furniss *et al.*, 1972).

The queen substance of the queen honey bee is instrumental in attracting a swarm outside the hive when colonies divide. Aggregation pheromones have also been demonstrated in ants and may occur in many insects that habitually form clusters.

Alarm pheromones are released when an intrusion or other disruptive and potentially threatening event occurs. Likewise trail pheromones occur in ants, termites and parasitic Hymenoptera.

Further, chemicals involved in inter specific communication are grouped as allomones or kairomones. An allomone is emitted by a member of one species and induces a response that is adaptively favourable to the emitter e.g. the repugnatorial substance released by a bug in response to a predator would be acting as an allomone.

A kairomone is adaptively favourable to the receiving organism e.g. the mosquitoes *Aedes aegypti* is attracted by lactic acid as well as other compounds in human perspiration. In this situation, the mosquito (i.e. receivers) benefits and hence the lactic acid is acting as a kairomone. In many ants, bees, wasps and termites the same chemical can function both as an allomone and pheromone e.g. in ants, formic acid can act as a defensive allomone and alarm pheromone that alerts nest mates to an intruder (Blum, 1978). Likewise, a volatile or mixture of volatiles may act as both, a pheromone or a kairomone. In Scolytidae (Coleoptera) e.g. the aggregating pherom-one may also act as to attract predators and parasites of the beetles.

At the end, it may be mentioned that death pheromones constitute another chemical messenger in social insects. Dead bees or ants emit a pheromone (oleic acid) that stimulates workers to remove their dead bodies from the nest.

18

INSECT PEST MANAGEMENT BY BEHAVIOURAL MEANS

An insects is considered a pest if it threatens a resource valued by humans, including human health (Foster and Harris, 1997). Protection of a resource from a pest is usually achieved by insecticides, but it can also be achieved by manipulating a behaviour of the pest. The manipulation of a pest's behaviour to protect a resource is not a new concept as rightly remarked by Foster and Harris. The practice of trap cropping i.e. using a sacrificial resource for the pest to attack, in order to protect a valued resource, has been known for centuries. However, in the last few decades largely due to improvements in analytical techniques and an increased desire to reduce the reliance on broad spectrum insecticides, there has been increased interest in the control of insect pest population by behavioural means. In this brief account, it is not possible to go in details but control of insect pests by behavioural means has been described which is based on Foster and Harris (1997).

BEHAVIOURAL MANIPULATION METHODS OF PEST MANAGEMENT

There are three main elements of a behavioural manipulation method : (1) a behaviour of pest, (2) a means by which the behaviour is manipulated appropriately, and (3) a method that utilizes the behavioural manipulation for protection of a resource from the pest. Critically speaking, any behaviour of any stage of the pest can be chosen, provided that its manipulation protects the resource. Intuitively, one might expect that manipulation of a pestilential behaviour (i.e. feeding on the source) or a behaviour closely related to the pestilential behaviour (i.e. finding the resource) is more likely to be useful for IPM than manipulation of behaviours unrelated to the resource e.g. mating.

The behaviour of an insect results from integration by its central nervous system of a variety of inputs that derive from stimuli acting on (a) exteroceptors, (which sense events external to the insect) (b) enteroceptors (which sense the internal physiological state of the insect) and (c) proprioceptors (which sense the relative positions of parts of the body).

The choice of a stimulus to use for behavioural manipulation is usually dependent upon a number of attributes including the following :

1. *Accessibility* – The stimulus must be suitable for presentation in a form that the insect can perceive.

2. *Definability and Reproducibility* – The more precisely that the stimulus can be defined, the more precisely it can be reproduced artificially.

3. *Controllability* –: The ability to control various parameters of a stimulus including intensity and longevity will give greater control in a behavioural manipulation.

4. *Specificity* – The more specific a stimulus is to a particular behaviour of a pest, the more likely it can be used to manipulate that behaviour. Conversely a stimulus that is ubiquitous in the environment is unlikely to be useful for manipulating specific behaviours unless its intensity or quality can be perceived by the insect above the background level, e.g. super normal visual targets (i.e. objects that reflect high ratios of stimulatory to inhibitory wave lengths of light), are used to outcompete the natural visual back-ground reflecting a lower ratio of stimulatory to inhibitory wave lengths.

5. *Practicability* – Environmental hazards and cost of protecting a resource must be within practical limits e.g. chemicals that are persistent and have high mammalian toxicities may protect an edible resource but render it useless for human consumption.

SOME EXAMPLES OF BEHAVIOURAL MANIPULATION METHODS

Stimuli that act over long distances

Attract - Annihilate

The attract annihilate (i.e. to attract and destroy) method is by far the most widely used behavioural manipulation for insect pest management. The strategy of the method is simple i.e. attract the pests to a site where as many of them as possible can be removed from the environment. The behavioural manipulation involves a long distance attractant and the mechanism consists of a device whereby the attracted pests are killed or trapped. This method has been used for many insect pests whose behaviour including long distance orientation, usually flight but also walking as in the case of certain species of cockroaches. The most commonly used attractants are volatile chemicals but visual stimuli are also used and auditory stimuli could also be used.

Chemical stimuli

Sex pheromones have been identified for a large number of insect pests, particularly Lepidoptera. These chemicals have a number of useful attributes for the attract-annihilate method including specificity, eliciting long distance responses and longevity in the field. However, because most sex pheromones are produced by females and elicit responses from males, they have been used primarily in the mating disruption method or for monitoring rather than in the attract-annihilate method. There are a few examples of effective control by mass trapping based on sex pheromones, including the citrus flower moth *Prays citri* on lemons in Israel, sporadic outbreaks of the gypsy moths in the United States and stored product pests in warehouses. Sex pheromones have also been used as attractants to facilitate contact with and dispersal of pathogens in pest populations.

The limitation of a sex pheromone attracting only males can be overcome by combining it with an attractant for females. Such a combination should be more effective in the attract-annihilate method than either attractant- alone. A combination of the sex pheromone which attracts males and a food lure (i.e. a mixture of phenethyl propionate, eugenol and geraniol) which predominantly attracts females, has been used against the Japanese beetle, *Popillia japonica*. The combination trapped more males and females than the two attractants did when used separately. Visual stimuli are also important for *P. japonica* and the catch of beetles is grater in white traps than those of other colour.

Aggregation pheromones attract both sexes and in some species immatures, e.g. nymphs of the German cockroach, *Blattella germanica*. Their ability to attract female makes these pheromones well suited for the attract-annihilate method. Aggregation pheromones have been used successfully for controlling various Coleoptera including the cotton boll weevil, *Anthonomus grandis* and the bark beetles in America and Europe.

A chemical need not be a natural stimulus in a behaviour to function as an attractant. Many so called moth sex attractants, which attract male moths but have not been identified in conspecific females, have been found by screening large numbers of related chemicals. A group of chemicals that has proven very effective as attractants for certain species of tephritid flies are the male lures. The most important of these chemicals are methyl eugenol, 1-(p-acetoxyphenyl)-buton 3-one (cue-lure) and t-butyl 4 (or 5) chloro 2-methyl-cyclohexanoate (trimedlure) which attract males of various species of Tephritidae of Old World tropics and subtropics and Oceania. Despite a great deal of investigation, these chemicals have no known role in the natural biology of the flies yet they elicit much greater responses than any natural sex pheromone in the respective species of flies (Foster and Harris, 1997).

Methyl eugenol was first used in a male eradication programme for the Oriental fruit fly, *Dacus dorsalis* by Steiner and Coworkers (1965) on the Rota Island. Essentially, they absorbed a mixture of methyl eugenol and insecticide (naled) onto cane-fibre squares and either threw them out of an aeroplane in uninhabited area or placed them in trees in village areas. This method, either alone or in combination with another such as sterile male technique, subsequently has been used a number of times to control or eradicate populations of Oriental fruitfly, *Dacus dorsalis* from several countries.

Visual and Chemical Stimuli

Visual stimuli are used most frequently in combination with chemical stimuli, enhancing the efficacy of a method over the use of either stimulus type alone. Even in the methods that do not mean to use visual stimuli, such as sex pheromone baited traps, the visual stimuli of traps are important. Tsetse fly (*Glossina* spp.) which are the vectors of the protozoans that cause trypanosomiases in Africa, including human sleeping sickness and important animal diseases, provide one of the best examples of the use of both visual and chemical stimuli in an attract-annihilate method. Attract-annihilate methods are particularly suitable for control of tsetse fly, because of the adenotrophic viviparity and low reproductive potential of each female, which commonly produces less than ten offspring. The host finding behaviour of tsetse fly is influenced by visual stimuli, including shape, orientation, brightness, contrast, movement and colour. Traps using only visual stimuli, such as the biconical trap, are used for control of these flies. Tsetse flies also respond to host odours, some of the attractive components of which include CO_2, 1-octen-3-ol, butanone, acetone and various phenols. The addition of odour (acetone and CO_2) to the biconical trap doubled catches over a non-odour baited trap.

In recent years, control of tsetse fly has relied increasingly on the use of odour baited black cloth targets coated with insecticide. The combination of visual and olfactory stimuli appears to concentrate a fly's movements toward the target, as well as to increase its chances of landing on it. Recently, the efficacy of odour-baited tree stumps with a nearby insecticide-treated netting was investigated for tsetse fly control, the method was effective with short trunks but less so with taller, upright trunks which tsetse flies avoid. An attract-annihilate target method utilizing both chemical and visual stimuli similar to that used for *Glossina* flies is being developed for screw worm control.

DISRUPTING BEHAVIOUR USING ATTRACTANTS AND REPELLENTS

Disrupting behaviour with stimuli that either elicit or inhibit orientation is also effective at long distances. In practice, only attractant and repellent chemicals have been used. The distinction between attractants and repellents is less clear than the names suggest. Most stimuli that attract will repel at higher concentration. Stimuli that repel a pest may elicit orientation in the same or other species. These terms, particularly ``repellent'' are often used problematically in the literature because the actual effect of the chemical is suspected from the end result of the behaviour, which could be influenced by short range stimuli such as deterrents. Nevertheless, attractants and repellents are generally identified as such in behavioural manipulation methods and we follow conventional usage.

Attractants

Most work on the use of attractants to disrupt a finding behaviour has focused on mate location particularly of moths in the so called mating disruption method. Large amounts of synthetic sex pheromone are distributed with the aim of preventing males from finding females. This method has been used successfully for the control of such herbivorous insect pests as the pink bollworm, *Pectinophora gossypiella* on cotton, the oriental fruit moth, *Grapholitha molesta*, on stone fruits; the tomato pinworm, *Keiferia lycopersicella*, on tomatoes etc.

In almost all cases where the method has been attempted disruption of mating has been at least partially achieved in the treated area. But it must be noted that the success of the method for management of a given pest depends to a large extent on its biology and particularly on the potential for immigration of mated females from outside the treated area.

A number of possible behavioural modes of action have been suggested for mating disruption with the help of synthetic pheromones including diminution of response due to sensory adaptation or habituation, arrestment of upwind flight at high concentrations, shifting the rhythm of response to females, changing the fine structure of or camouflaging a natural plume, outcompeting females and causing an imbalance of sensory inputs by altering the perceived blend.

An interesting variant of mating disruption method is used for control of the pink boll worm in California: an insecticide `per methrin' is added to the sticker used to attach the pheromone formulation to the leaves of the plant. This combination is thought to increase the control of bollworm over the pheromone alone.

Repellents

The strategy for using repellents as remarked by Foster and Harris (1997) is generally to stop a pest from finding a valued resource. Useful repellents can be derived from natural sources such as insects e.g. defence secretions or they may be purely artificial as in the case of most insecticides. Most work on the practical use of volatile repellents has been to protect humans from insect bites e.g. mosquitoes and black flies that are vectors of diseases. Of these chemicals, the best known is ``deet" which is used as a repellent for a wide range of insects. Repellent like pine oil has been used to prevent oviposition by the onion fly, *Delia antiqua*.

Stimulants

Most known stimulants are involved in either feeding or oviposition, primarily the former. Advantages of stimulants including increasing exposure to toxins that must be ingested and applicability to a wide range of insect pests, as in the case of sucrose.

Feeding stimulants also are often common carbohydrates, proteins or fats that are easily obtained and relatively inexpensive, whereas oviposition stimulatns can be highly specific, even among pests that threaten same resources and expensive. Further, feeding stimulants are specially useful in conjunction with toxins because they can increase contact that may be suppressed if the pest responds to the toxins by ceasing to feed, thus avoiding a lethal dose e.g. endotoxins of *Bacillus thuringiensis*. They have also been used in combination with attractants (attract-annihilate method) using commercially available stimulants, e.g. protein - hydrolysate in the eradication of tephritid fruit flies and sugars for control of flies around animal rearing facilities.

Deterrents

A deterrent is a chemical that inhibits behaviour, such as feeding or oviposition when applied to a site where such behaviour normally occurs. In pest management a deterrent is applied directly to a resource to prevent or reduce the effects of a pestilential behaviour such as feeding. Efficacy depends on the physiological state and behavioural responses of the pest during initial and subsequent encounters with the deterrent. Desensitization is a direct effect of deterrent. Numerous insect pests have been shown to lose their responsiveness to deterrents after repeated exposure and after increasingly long periods of deprivation of feeding or ovipositional sites.

Deterrents found in the extracts of Neem tree, *Azadirachta indica* have attracted great interest in recent years specially azadirachtin. Neem extracts and azadirachtin affect behaviour, growth regulation, ovarian development, fecundity and fertility in insects. Neem chemicals may control insect pests that will not consume them by deterrent effects. Lepidoptera are most sensitive to azadirachtin with feeding deterred at $< 1\text{-}50$ ppm. The dual action of deterrence and toxicity and the many pest species that are controlled by one or both of these actions contribute to interest in Neem chemicals.

Only a few plant derived feeding deterrents in addition to Neem have been tested in small glasshouse or field plots. Polygodial, a plant derived dialdehydic sesquiterpenoid in the drimane series, is a feeding deterrent for various lepidopteran pests such as *Spodoptera* and *Heliothis* species and for the aphid *Myzus persicae*. In the case of *M. persicae*, deterrence of feeding decreased spread of persistent and semi-persistent plant viruses. The deterrent properties of fungicides have been known for many decades and were discovered through observations of pest populations feeding on treated and un-treated plants.

In addition, use of attractant, repellents and deterrent as well as stimulants in combination of any two or more increases the efficacy of the chemicals as suggested by several workers.

Thus, it can be safely stated that behavioural manipulation methods for pest management can be developed in a variety of ways and the behaviour of insects is influenced by many stimuli both external as well as internal. Internal stimuli related to different physiological and experimental stages are important sources of behavioural variability and should not be ignored as pointed out by Foster and Harris (1997). There have been a limited number of successful examples of behavioural manipulation methods in pest management but further research work on the above field will definitely prove the significance of behavioural means in controlling the insect pest population.

19

MIGRATORY AND DISPERSAL BEHAVIOUR OF SOME INSECTS

It is a matter of common experience that during spring when the sky is overcast by clouds or the temperature is most suitable, many species of winged aphids migrate from one place to another in thousands. Many such spring migrant aphids cause a serious havoc to the Leguminous and Cruciferous crops and the flowers of a variety of trees which finally cause a considerable loss of fruit yield.

Likewise, in winter the swarms of locust, *Schistocerca gregaria* migrate from Africa to India via Pakistan and during their migration the swarm composed of several thousand locusts in the sky appearing like a dense cloud; causing a lot of damage to almost all sorts of plant species *en route,* since they are polyphagous. The locusts are normally diurnal i.e. active during day time in habit and the swarm flights have been observed up to dusk in cold weather i.e. winter (the temperature being below 4°C) as observed by the author in the Rajasthan desert. They rest during night hours on any plant and till early morning they become so inactive and motionless due to cold that one can collect bags of locust before sun rises. It has been observed that just by shaking the trees and bushes, on which they rest at night thousands of locusts fell on the ground and they can be easily collected in large bags and big containers. After the sun rise, temperature gradually increases and the locusts swarm again become active and start flying *en masse* in any particular direction for which the direction of wind current and solar radiation play a very vital role.

The migratory behaviour of almost all micro-and mega-insects depends on temperature, humidity, intensity of light and wind velocity. All these factors work together and one can not single out any of the above mentioned factors. The moths and butterflies also migrate from one place to another for the want of food, shelter, breeding place etc. while the moths migrate during dusk and night, the butterflies migrate at day time. In a similar manner many insects like aphids, white flies, dragon flies, several heteropteran insect pests have been observed migrating for the want of favourable life conditions (food, shelter and breeding place).

In the following pages a brief account on this subject i.e. migratory behaviour of some insects is given. However, readers interested for having a comprehensive account on almost all aspects of insect migration, not dealt here, are suggested to consult the book by Johnson (1969) and a recent review of Gatehouse (1997), which forms the basis of this article.

PART-A

MIGRATORY BEHAVIOUR OF SOME WIND BORNE INSECTS

The phenomenon of air or wind borne migration, in which migrants ascend out of their flight boudary layer to be trasnported down wind has evolved independently in several insect orders. Glick (1939) recorded representatives of 18 insect orders in day time aircraft net catches at altitudes from 60 m. to 1600 m. but only 12 orders were recorded during a less extensive programme of night-time trapping. According to him the wind borne insect migrants range in size from minute flies to large grasshoppers, locusts and moths, but adaptive wind borne migration has been most intensively studied on the Homoptera (aphids and white flies), Orthoptera (locusts and grasshoppers) and Lepidoptera (moths and butterflies).

It has been reported that some nocturnal migrants extend the scale of displacements they can achieve by flying on more than one successive night. The day flying migrants, which may be adapted to tracking changes in the availability of their habitats on a more restricted spatial scale e.g. aphids, which appear typically to make only a single migratory flight, though this may persist into the following night resulting in very substantial displacements. For those insects, which are nocturnal migrants, the altitude reached is determined by the behaviour of the insects. Short flight duration may limit some migratory insects to levels at which air temperatures are still well above their thresholds for flight as was observed in the case of cecadellids. A common feature of the nocturnal migrants e.g. grasshoppers and moths is their concentration in layers well below their temperature determined flight ceiling.

The frequent occurrence of layer concentrations of nocturnal migrants and their association with recognizable atmospheric conditions as well as a high incidence of collective orientation, suggests that they must generally be the outcome of adaptive behaviour.

Another conspicuous and frequent feature of the behaviour of many nocturnal migrants revealed by radar, is mutual alignment achieved by common orientation. Orientation has been observed in larger migrants like locusts, and moths.

Concentration and Descent Behaviour

Wind borne migratory insects are subject to the effects of convergent wind systems, on the micro, meso and synoptic scale, which are capable of increasing their aërial densities up to several orders of magnitude. Such concentration may result in high densities of reproducing insects and thus larval outbreaks when it occurs as migratory behaviour ends. This is more likely to happen following concentration in slow moving or stationary zones of convergence. The ecological significance for wind borne migratory insects of accumulation in large scale convergence systems e.g. the inter tropical convergence zone and warm ``conveyer belt" winds in front of cold fronts in certain temperate regions appears to lie in drawing insects into regions with an increased probability of rainfall. Less is understood of the ecological significance of concentration in a smaller scale systems, although intense convergence associated with convective storm outflows is known to draw migrating *Spodoptera exempta* moths into areas of recent rainfall, which provide larval habitats.

It has been observed that in early evening the air below a developing temperature inversion rapidly empties of diurnal wind borne migrants as air temperature fall below the threshold for flight and take off is inhibited by low light intensity. However, insects that are still at altitude may become concentrated in the warm air at the top of the temperature inversion, they remain airborne and travel rapidly downwind. When nighttime air temperatures below the temperature inversion remain above threshold, flight by diurnal insects may also persist into the night.

A general descent of nocturnal wind borne migratory insects at, or shortly after, dawn is often observed in Radar studies. However, nocturnal flights by *Nilaparvata lugens* may persist after dawn and individuals taking off at dawn sometimes reach high altitudes and form layer concentrations that last for up to 3 hours. In specific weather conditions, aerial trapping has shown that migration by this species at high densities may continue into the afternoon.

Although light rain may not affect much to the migrants but heavier rainfall can cause the migratory insects to descend. It has been found that the Radar and optical detectors as well as direct observation, provide good evidence that moths descend to fly within 1-2 m of the ground or settle when they encounter rainfall. Further, it may be mentioned that, whether or not migratory insects are washed out by rainfall in this way are responsible for initiating larval outbreaks is likely to depend on whether they have terminated migratory flight and are reproductively mature at the time.

Benefits of wind borne Migration

Wind borne migrants become dispersed as they are carried on the wind and as migratory flight terminates and they descend, their destinations are distributed across the landscape down wind of their source. Wind borne migration can only evolve where there is a high probability that suitable habitats occur downwind of the point of departure, so that as they reach their destinations, the migrants sample the landscape for these habitat patches. The spatial extent and density of the distribution of their destinations depend on the distance travelled by, and degree of dispersal of the migrants in flight. Displacement and dispersal depend in turn on the speed and variation in direction of the winds the insects encounter and on their behaviour.

It must be pointed out here that Kennedy's early (1961) recognition of the distinctive characteristics of migratory behaviour has culminated in a definition and now the widely accepted view that migration is best regarded as a behavioural process, that with movements associated with foraging and adventitious and accidental movements, influences the spatial dynamics of insect populations.

Termination and Reproductive Maturation

The evidence now available suggests that migration occurs pre-reproductively i.e. before first oviposition or larviposition in most migratory species representing a large number of insect orders. In some of these species e.g. aphids there is evidence also of interreproductive migration i.e. migration between attempts of oviposition or larviposition, but for many migration appears to be exclusively pre-reproductive. Only for the spruce budworm *Choristoneura fumiferana* is there good evidence that migration is exclusively interreproductive.

In spite of recent claims to the contrary, migration in noctuid moths appears to be exclusively pre-reproductive, e.g. the migratory behaviour persists after reproductive maturation and mating in these moths is often observed fact that the first female immigrants trapped in an area are found, almost without exception, to be mature and mated. However, if migratory behaviour were to persist after maturation and mating, a significant proportion of migrating insects captured during migratory flights, might be expected to be mature and mated. In fact, the overwhelming majority are found to be immature and unmated, and special circumstances appear to explain the few exceptions to this observation e.g. a relatively high proportion of female *Helicoverpa zea* caught on oil rigs in the Gulf of Mexico were found to be mated. The same was true of a small sample of *Mythimna separata* captured in a kite-borne trap at 200-300 m. in China, having crossed the Bohai Sea. In both cases, the insects may have been constrained to continue flight over the sea. Mature mated females of *H. zea* were also caught about 150 m. on a television tower in Georgia.

Recent data on the effect of maturation on tethered (limited) flight behaviour by *S. exempta* clarify the relationship between the termination of migratory behaviour and reproductive maturation. Although there is evidence of a decline in tethered flight activity associated with reproductive maturation for several migratory Hemiptera and Lepidoptera including *S. exempta* and for *Melanoplus sanguinipes,* recent experiments have examined the precise temporal relationship between the onset of calling (pheromone release) and tethered flight behaviour in female *S. exempta* (see : Gatehouse, 1997).

Post-Migratory Behaviour

The migration of insects presumably descend as migratory behaviour ends with the disinhibition of station keeping responses. Nothing is known of the behaviour involved in descent except in those Homoptera for which the reappearance of responsiveness to longer wavelength light, reflected from plants have been shown to characterize the end of migratory behaviour e.g. aphids.

Observations of the flight behaviour of *Helicoverpa* and *Heliothis* species in the field indicate that these moths are highly vagile and undertake extensive flights during reproduction. The observations suggest that their post-migratory flights, after migratory behaviour has terminated outside the suitable habitats, may also be extensive.

PART - B

THE MIGRATORY AND DISPERSAL BEHAVIOUR OF SOME INSECTS

There is no doubt that a specific phenomenon generally called migration (vide supra) exists among insects. It is usually exemplified in the transfer of population from place to place by mass flights.

Fig. 63A. Migration of Monarch Butterflies

For many Entomologists a true migration happens only when a crowd of insects travel together, more or less in the same direction which they are thought to control and when after a sojourn at the destination the same individuals return more or less to the former habitat much like swallows. The Monarch butterfly, *Danaus plexippus* (Fig. 63A) is an example, it migrates from Canada in the autumn to over winter sites in California and Florida, and some of the same individuals return north again in following spring and breed. The ``return" by the same individuals has been called ``re-migration".

Fig. 63B. After migration, millions of Monarch butterflies are seen resting on suitable sites of Mexican uplands (see text).

Further, according to UNESCO report (Anonymous, 2009) the Mariposa Monarca Biosphere Reserve is located about 100 km (62 miles) north-west of Mexico City. Every year it becomes the home of millions of monarch butterflies which seek out the moderate climate of the Mexican uplands for their winter quarters.

The biosphere reserve covers 56,259 ha (139,000 acres) of montane forest at an elevation of about 3,000 m (9,900 feet) above sea level, running along the border from the state of Michoacan to the capital, Mexico City. It is named after the migratory monarch butterfly, which in the fall travels south for some 4,000 km (2,500 miles) from the USA and Canada to Mexico. Every year the bare rocks of the Mexican plain are the scene of a fascinating natural spectacle as several hundred million monarch butterflies arrive on the reserve and turn the whole landscape orange (Fig. 63B). The butterfly's migration is an astounding phenomenon; hatching in North America, the creatures come to Mexico, passing the winter in a kind of suspended animation. In the spring they return north, crossing the southern states of the USA and depositing their eggs on the leaves of the milk weed plant, a poisonous plant with silky leaves, whereupon the generation that has just spent the winter in Mexico dies. After metamorphosis, the new butterflies continue the journey north, and during the summer months three further generations will pass, some of whom may begin the journey south again. Only the fourth generation will return to Mexico. The biosphere reserve was established in the 1980s and houses several *sanctuarios,* or protected zones, intended to preserve the butterfly's habitat from increasing human settlement and unregulated logging.

In most species the adult insects do not live long enough to return at a later season and although many species have individuals that periodically fly together in the direction with the persistence that is generally regarded as characteristic of migrants, such insects may make only the one way journey, lay eggs and die instead of returning as in the case of desert locusts *S. gregaria* (Fig. 63C), aphids (Fig. 63D), white flies, termites etc. It is their progeny that recolonize the former habitat and this process also has been called remigration. Williams (1930, 1957, 1958), who was particularly concerned with the migration of the Lepidoptera recognized this one-way travel and from him, as for many other Entomologists (e.g. Nielsen, 1961), a migration need by only a one-way journey provided it is a continued movement in more or less definite direction, in which both movement and direction are under the control of the animal concerned (Williams, 1958) and which results in a temporary or permanent change of habitat. Nevertheless, Williams persistently looked for two way journeys among such migrants for he believed that migration that was one way only would end in the extinction of the species, an apparently inconceivable result for such a firmly established process. He considered that more and more two way journeys were being proved. Severin (1933) working with leaf hoppers that migrate on the wind, also regarded migration as a fatal journey beyond the habitat. Movement within the habitat he considered to be different and he called it ``dispersal''.

Fig. 63C. Part of swarm of Locusts

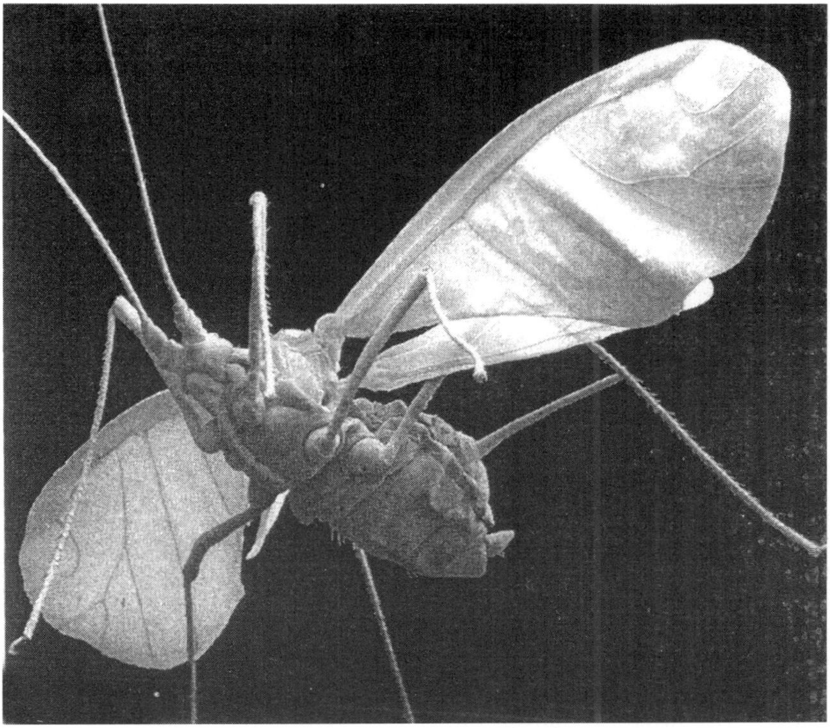

Fig. 63D. Windborne Aphid

Workers with aphids nevertheless speak of migratory species namely those whose individuals are carried seasonally on the wind between primary and secondary hosts, in distinction to non-migratory species that breed only on one kind of host although their alatae may be transported by flight for hundreds of kilometers.

Undoubtedly many insects are blown about accidentally or inadvertently. But many species both strong and weak fliers are also specifically adapted to travel on air currents. The best known of all migrants, the strongly flying locusts, once thought to control the direction of their long distance flights are now known to travel on the wind in a manner that has become adapted for the locusts to reach new seasonal breeding places thousands of kms. away from the old ones; it now seems that many well recognized migrants including some Lepidoptera (moths as well as butterflies) that were commonly thought to control their direction also travel on the wind adaptability.

Ecological Aspects of Migratory Behaviour

Many authors stress that migration involves a departure from the place, locality habitat, field of operation or territory and that it results in the migrants crossing places where normal life is not possible (see: Johnson, 1969). It is important to realize that this place is primarily the one where the adults emerged or one very close by. Unless this is stated, migratory and trivial flights for example those from flower to flower by a butterfly, are not distinguished. To this distinction must be added the qualities of undistractedness and lengthly flight that take the insect well beyond the place of emergence to another site where the breeding can continue.

Migration is often linked with the need of a population to relinquish a habitat that is only transiently fit for breeding. Migration is, therefore, not only a peculiar type of flight behaviour. It is also flight that is adapted for periodically transporting insects beyond the boundaries of their old breeding habitats and into new ones. This involves a stage when adults leave the breeding habitat, a stage of transport and a stage of habitat-finding and acceptance, these stages differing from order to order, species to species and time to time. The concomitant behaviour differs even from one individual to another in the same species according to the boundaries of the breeding habitats which may be very localized or many extend across a continent. Some flights last only a few minutes and take the insects only a few metres as with ants and termites, others last for hours or intermittently for weeks while the insects travel hundreds or thousands of kilometres, as with locusts.

Individual and Collective Aspects of Migratory Behaviour

Individual flight behaviour is one aspect of migration whereas the collective ecological system which produces migrants and is produced by them is another. Individual physiology and behaviour and collective function are integrated and inseparable in nature and only for analysis can they be considered apart.

Large populations, specially the phytophagous insects tend to diminish the quanity and quality of food. Apparently this sometimes causes larvae to develop into sexually immature adults with power of migration instead of into sexually mature insects that do not migrate. Thus, the character of individuals is affected by the community in which they live and which in turn they influence.

The breeding habitat is a place where the eggs are laid, larvae develop and adults emerge. It is a plant or tree for aphids, an area or ground with vegetation for locusts and grasshoppers, a pond, ditch or marsh for mosquitoes and carrion and the soil beneath for blowflies, dung for houseflies, rotten meat for flesh flies etc.

Most flights recognizable as "migratory" begin either where the adult migrants emerge or close by and they may end where migrating insects lay their eggs. Migration is essentially a transference of adults of a few generation from one breeding habitat to other.

Clearly, the long migratory flights of the desert locust, *S. gregaria* are adaptive and have become geared by selection to take the insects from places that become useless for breeding, owing to seasonal drought to others where seasonal rains permit breeding to continue. These flights have now become an integral part of the life of the species.

The flight apparatus which includes wings, wing-muscles and the biochemical process associated with their functioning as well as the storage and use of fuel for flight is extremely labile in its development. The whole apparatus for flight may develop to maturity or its development may be inhibited at any stage, wings may be short or muscles may remain incomplete. Threshold of neural response associated with flight also vary according to the stage of development of the insect.

The development of two systems, ovaries and flight apparatus is affecterd by environmental factors and specially by food, crowding and temperature. If the flight apparatus becomes fully mature while the ovaries remain immature, the insect may become migrant. If ovaries become mature and the flight apparatus does not, eggs are laid in or near the old habitat. Variations in structure and behaviour between the two extremes, migrant adult and flightless adult are geared to the ecological needs of the species.

As stated, migratory flight in insects is evoked not capriciously but at a definite stage in the life of the adult. Adult insects may pass through several stages of immaturity and maturity between emergence and death and their behaviour, particularly flight behaviour varies accordingly. In most of the literature on migration and dispersal the migratory behaviour of the two sexes is not distinguished perhaps because both sexes of some will known migrants such as desert locust and many butterflies, migrate together for long distances. With many species the sexes differ greatly in their migratory behaviour though it is usually the females that migrate, males being more variable in their performance. In many of the records listed below no distinction was made between the sexes and sometimes the behaviour of the males might be uncertain.

Orders of Insects that Show Migratory Behaviour

1. Odonata : Dragon flies viz. *Libellula quadrimaculata,* L. *depressa, Leucorrhinia pectoralis,* L. *rubicunda, Aeschna juncea* etc.
2. Orthoptera : The grasshoppers viz. *Melanoplus sanguinipes, Dissosteira longipennis* and locusts : *Schistocerca gregaria* and the red locust *Nomadacris septemfasciata* etc.
3. Isoptera : Termites usually not regarded as migrants.
4. Homoptera : Aphids, viz. - *Aphis fabae, Myzus persicae* etc.
5. Heteroptera : Bugs - *Dysdercus fasciatus, Dysdercus nigrofasciatus, Dysdercus superstitiosus, Dysdercus koenigii,* etc.
6. Lepidoptera : Butterflies viz. *Vanessa cardui, Ascia monuste, Pieris brassicae, Nymphalis io, Danaus plexippus.*

   Moths - *Choristoneura fumiferana, Porosagrotis orthogonia* etc.
7. Coleoptera : Beetles viz. *Lucanus cervus, Melolontha melolontha, Anthonomus grandis, Bruchus pisorum, Dendroctonus frontalis, Leptinotarsa decemlineata* etc.
8. Diptera : Mosquitoes viz. *Aedes sollicitans, Aedes taeniorhynchus, Anopheles maculipennis, Anopheles annulipes.*
9. Hymenoptera : Ants viz. *Myrmica emeryana, Formica obscuriventris* and *Gilpinia polytoma* etc.

In the following pages only a brief account of the migration and migratory behaviour of butterflies, locusts, bugs and aphids is given (For details see : Johnson, 1969).

MIGRATORY AND NON-MIGRATORY FLIGHT

The flights of a migrating butterfly differs so much from other types of flight that even a casual observer is able to identify a migrant. In describing the migration of *Ascia monuste,* Nielsen (1961) noted that the individuals do not usually stop for food, although occasionally they may feed hurriedly before continuing to migrate always in the same direction as formerly and independently of the wind direction. According to Nielsen the same course is resumed even after a night's rest so that up to 160 km. may be traversed by

A. monuste is more or less continuous flight, before it ends its migration and its flights become shorter, more local and in all directions.

For the Monarch butterfly, *Danaus plexippus,* the wings during cruising are said to bear with a restricted amplitude of about 30° but a significant feature of the flight is its persistence day after day, although the insect rests at night. Many marked Monarchs have been recaptured up to nearly 3200 kms from where they were bred and librated; moreover, such estimates make no allowance for deviations from a straight track.

The persistence differs greatly with other types of insect e.g. desert locust *S. gregaria* keep flying off and on more or less every day for many weeks and traverse hundred or even thousands of kms. to a new breeding area, yet they land frequently beneath the swarm, taking off again as the swarm passes over them and feeding voraciously when the swarm descends to rest in the evening.

Some Miridae (bugs) make relatively long dispersal flights from the bushes in which they were bred, only when they are young adults and they spend the rest of their lives making flitters within and round the habitat. A new born alate black bean aphid, *Aphis fabae* usually rejects the perfectly suitable plant on which it was bred and takes off almost irresitibly on a flight which effectively removes it from its birth place.

It has been found that when alternative breeding sites or hibernation sites are normally far away from the original breeding sites, migrations are prolonged as with *S. gregaria* or *D. plexippus*. When habitats are scattered over a breeding range, individuals migrate and oviposit (or larviposit) alternately and repeatedly e.g. aphids and flesh flies and the periods when the insects are undistracted by stimuli that would normally divert them to habitats or mating may then be very short.

With such insects as migratory aphids, desert locusts and migrating butterflies, it is not difficult to distinguish flight that is migratory from other kinds of flight. This is because flight is particularly persistent and rectilinear as with *Ascia monuste* or as with *S. gregaria* most of the active flight occurs while the adult insects travel *en masse* between seasonal breeding grounds or alternatively because flight occupies a very brief period of adult life and is associated with no activity except movement between seasonal breeding habitats as with some parthenogenetic migratory aphids. The criteria that distinguish migratory flight are thus varied but have the common factor of relative persistence and undistractability from normal stimuli that in other circumstances would cause flight to cease.

Seasonal Movement of the Desert Locust in Relation to Seasonal Winds

The desert locust, *S. gregaria* invades a total area of about 17 million sq. kms. in Africa and South West Asia and extends even into Europe. But it breeds over only abut 46% of this invasion area or over rather more than 8 million sq. kms. The breeding and migration cycle within and between the different parts of the whole invasion area is extremely complex and only a broad and oversimplified account is possible. Most of the invasion area is arid or semi-arid and the locusts breed only where seasonal rains allow and because these rains and the associated breeding occur at different times of the year in different parts of the invasion area, locusts must migrate from one breeding area to another.

There are two great zones within which the locusts breed in alternate halves of the year, and the cycle of breeding and migration has become adapted so that swarms are displaced between one zone and another on appropriate seasonal winds. The extent of the breeding areas within these zones varies annually and some areas may escape invasion for several successive years.

Spring breeding occurs within a zone that stretches from north western and northern Africa through the middle East to Pakistan and through the Somali peninsula to East Africa, where it is associated with the cyclonic rains characteristic of the Mediterranean area and with the local seasonal rains in the Red sea and East African regions, respectively.

In the second half of the year, locusts breed in a zone associated with rains at the inter tropical convergence zone which contains an area of summer breeding and one of winter breeding. The summer breeding area stretches south of the Sahara from Senegal and Southern Mauretania across Africa to Chad and Sudan republic, Ethiopia, Southern Arabia, India and Pakistan. The main areas of winter breeding are in Somali peninsula and southern Red Sea basin.

Swarms of young locusts fly out of the seasonal breeding areas and move with the wind to the next seasonal breeding area. Most parts of the invasion area regularly become clear of locusts for periods of months and some areas may have no locust for years. Emigrating swarms seldom breed within 1000 kms. of their sources and often traverse 2000-3000 kms. in one generation. In this way all parts of the total invasion area where breeding occurs are linked either by swarms flying direct from one part to another or by swarms of one generation flying to an intervening breeding place and producing another generation that flies farther on at a later date. Individual swarms may move for more than 5000 kms. within two months as from the Arabian peninsula to Mauretania in 1950 or over 3000 kms. in less than four weeks, as when swarms flew from Somali peninsula to the borders of Jordan and Iraq in 1951-52 (Johnson, 1969).

In the last and first quarters of the year (October-March) swarms from both summer and winter generations invade areas where they will breed in the spring, during this period locusts may be very widespread. In the second quarter (April-June) swarms of old locusts reach the limit of dispersion and swarms of young ones of the spring generation appear and begin to spread back to summer breeding areas.

The swarms tend to become most concentrated during the third quarter (July-September) when they occupy a relatively narrow area across Africa from Senegal to the Red Sea and through the southern Somali peninsula, southern Arabia, Pakistan and India. The concentration is most evident in August and September when the whole of the north western Africa, northern Arabia and Middle east are usually free from swarms and hoppers. The pattern of these movements in relation to seasonal winds is extremely complex (see : Johnson, 1969).

20

INFLUENCE OF HOST PLANTS ON SEX PHEROMONE BEHAVIOUR OF PHYTOPHAGOUS INSECTS

The phytophagous insects often meet, court, and mate on plants and their sexual behaviour may occur principally or exclusively on host plants. Thornhill and Alcock (1983) noted that most sexual encounter sites are focused on resources attractive to females. For a phytophagous species, host plants are an appropriately attractive resource.

The phytophagous insects may mate on host plants because of the relatively high probability of chance meetings with the opposite sex in a place of mutual interest, particularly for species whose adults feed on host plants. Males may purposefully search host plants for females simply because females are likely to visit feeding and oviposition sites, resulting in mixed sex encounters and mating opportunities on host plants. However, there are numerous documented examples of specific relationships between insect sexual behaviour and host plants that indicate much more is occurring than the use of plants as favoured encounter sites. In many species of phytophagous insects, it appears that observed sexual activities on host plants often are the result of mate-finding and mate-selection strategies that integrate reproduction and host plants in a variety of ways (Landolt and Phillips, 1997). In the following pages a brief account on the subject is given which is based on the work of Landolt and Phillips (1997).

PHEROMONE PRODUCTION AND RELEASE

Sex pheromone production and release may be affected by the presence of host plant material in a number of different ways, both direct and indirect. Host plants undoubtedly play an important role in the origin of sex pheromones of some phytophagous insects through the acquisition of bioactive chemicals and required chemical precursors of pheromones via consumption, absorption, or inhalation of host plant material. In some species, pheromones are passed and released with faeces, providing a strong relationship between both host plant chemistry and pheromones. In these species, there may be a temporal and spatial relationship between feeding and pheromone release. Plant Kairomones may also act on the nervous and hormonal systems of insects to stimulate pheromone production and release in both males and females.

LARVAL SEQUESTRATION OF PLANT CHEMICALS

The most direct examples of a host plant role in insect sex pheromone production are of larval sequestration of plant chemicals and their subsequent use in sexual communication as adults.

Studies on several genera of arctiid moths and on certain groups of nymphalid butterflies clearly demonstrated that secondary plant compounds consumed by larvae are used by males either in sex attraction or courtship interactions. *Utetheisa ornatrix* males (Arctiidae) possess a courtship pheromone, ``hydroxydanaidal", that is derived from pyrrolizidine alkaloids (PAs) ingested in the larval stage from host plants of the genus *Crotalaria.* The compounds are sequestered as a defensive ploy by larvae and endow similar protection on eggs and adults. Males advertize their alkloid load by signalling with hydroxydanaidal, providing females a means of discriminating between males possessing the alkaloid and males without it. The plant-derived alkaloid is transferable to some degree from the male to the female via the spermatophore. This is a clear example of a sex pheromone precursor acquired from a host plant sequestered by the larva.

ADULT ACQUISITION OF PLANT COMPOUNDS

Males of the arctiid moth *Cisseps fulvicollis* appear to obtain pyrrolizidine alkaloids (PAs) from plants and use them as precursors to produce a sex pheromone. In this species, adults acquire PAs when they feed on dead and damaged plants that contain these compounds. The same pyrrolizidine alkaloids apparently attract males to plants and attract females to males. Males also produce *hydroxydanaidal* from these PAs in eversible coremata and release it as a sex attractant. Similar behaviour occurs in nymphalid butterflies. Ithomiine butterflies are attracted to and feed at plants containing PAs and sequester compounds as precursors of ``hairpencil'' pheromones. The hairpencil pheromones function in sex recognition and possibly in other courtship roles. Some adult danaiine butterflies also feed on plants containing alkaloids, and they use these alkaloids as biochemical precursors of courtship pheromones that are used to dust females upon approach.

Collection of chemicals from plants and their use in a sexual function is also known for certain species of euglossine bees. Male bees collect chemicals from orchids by scraping the plant's waxy cuticle and storing the material in pouches on the hind tibia. The material, a mixture of terpenoid compounds, is used either as an aggregation pheromone to promote the formation of leks or as a sex attractant that also attracts opportunistic males.

Some species of tephritid fruit flies also appear to acquire sex pheromone precursors from plants. *Bactrocera dorsalis* males are attracted to and feed on methyl eugenol, a substance encountered in some flowers. Males that have eaten methyl eugenol are more successful in courting and mating females than males that have not. Metabolic biproducts of plant acquired methyl eugenol appear to make up part of the pheromone produced by *B. dorsalis* in the hindgut.

Many species of beetles, primarily scolytids, form feeding and mating aggregations on host plants, as a result of attraction to sex pheromones. The often close relationship between feeding and sex attraction suggests that many examples may involve acquisition of plant compounds and their use as pheromones or pheromone precursors.

Sex pheromones of beetles at such feeding and mating sites may be plant derived due to (a) adult feeding on plant tissue, passage of chemicals through the gut, and release in the frass; (b) production in the gut from plant derived precursors; or (c) microbial action on plant material in the gut. Examples of demonstrated association between sexual attractiveness and food include the grain beetles *Sitophilus oryzae, Tribolium castaneum* and *Rhizopertha dominica.* In these species, pheromone is released only while males are feeding on grain, and there is no pheromone released when food is removed.

Certain species of scolytid bark beetles are known to produce terpenoid pheromones from the host tree terpenes. Males of *Ips paraconfusus* use myrcene from hosts to manufacture the attractive pheromones ``ipsenol" and ``ipsdienol", which are released by males after gallery formation in the host. For males of *I. paraconfusus* and *Ips avulsus*, exposure to vapours provides sufficient myrcene for pheromone production, but *Ips calligraphus* and *Ips grandicollis* males must feed before myrcene is converted to ipsdienol. Males of *I. paraconfusus* also produce the pheromone ``cis-verbenol" from the host compound (-) α-pinene. Other species may sequester myrcene as larvae in addition to acquiring myrcene as feeding adults. *Dendroctonus ponderosae* males use α-pinene from host trees as a precursor for producing the pheromone *trans*-verbenol.

PLANT STIMULATION OF PHEROMONE PRODUCTION

The stimulation by host plants of pheromone biosynthesis by insects has been reported for females of moths in the genus *Helicoverpa* and scolytid beetles. Production of sex pheromone by wild female corn earworm moths, *Helicoverpa zea*, is induced by the presence of host plants. Volatile chemical signals from corn silk, from solvent extracts of corn silk, or from tomatoes triggered the production of sex pheromone in terminal abdominal glands of *H. zea*. Moths removed from any host plant materials inhibit reproductive activity including pheromone production. In the bark beetle, *Dendroctonus brevicomis,* male pheromone synthesis is stimulated by feeding on the host tree. Feeding may trigger pheromone synthesis by removing inhibition of the corpora allata and triggering release of juvenile hormone. Juvenile hormone activates brain neurosecretory cells to release a stimulatory brain hormone that drives pheromone synthesis.

Host plant stimulation of pheromone release occur also in Lepidoptera. Females of four species of ermine moths (*Yponomeuta* spp.) are stimulated by their respective host plants to release sex pheromone attractive to males. In these species, females appear to be attracted first to their host plants, on which they then release sex pheromone. Males of the cabbage looper, *Trichoplusia ni*, release greater amounts of the pheromone component ``d-linalool" when exposed to the combination of odour of cabbage foliage and the female pheromone Z-7 dodecenyl acetate, than when exposed to the female pheromone alone. Female moths of this species were stimulated to release pheromone by the odour of cotton foliage. When airflow from over cotton foliage was passed over females, the airflow was more attractive to males than airflow from over females that was only subsequently passed over cotton foliage.

Females of the pink bollworm scent-mark cotton leaves are leaving pheromone that is attractive to males. Similar behaviour has been observed in male cabbage looper moths. Sex pheromones may also be adsorbed and rereleased from leaf surfaces. The pheromone of female *Mamestra brassicae* may be adsorbed onto host leaves in quantities sufficient to be attractive to males as well as to egg parasites.

MALE-PRODUCED PHEROMONES

Host odour enhancement of attraction responses to male pheromone occurs in several insect orders. In the Diptera, such effects have been demonstrated in Tephritidae and Drosophilidae. For example, the combination of male pheromone and host fruit odour is more attractive to female papaya fruit flies. *Toxotrypana curvicauda*, than is either male pheromone or host fruit odour alone.

Many species of *Drosophila* fruits flies are attracted to male pheromone in combination with food odours emanating from damaged host plant tissue. These odours may in part emanate from microbial colonies on damaged plant tissue rather than the plant itself. Attraction of female *Drosophila borealis* and *Drosophila littoralis* to male pheromone (whether that emanating from live males of the synthetic pheromone ethyl tiglate) is synergized by odour or fermenting bark of aspen, which is a host for oviposition and larval development.

Host odour enhancement of pheromonal response is known also for beetle attraction to sex pheromones. In the pine weevil *Pissodes nemorensis,* the male produced pheromones ``grandisol" and ``grandisal" were attractive in the field only when deployed with odours from a cut pine bolt. In the maize weevil *Sitophilus zeamais,* weevil responded significantly more to male-produced pheromons deployed with grain than to either pheromone or grain separately.

Male and female sap beetles (Nitidulidae) of the genus *Carpophilus* (dried-fruit beetles) are attracted to the combination of male pheromone and odours from damaged host fruit that is likely to emanate in part from yeast colonies on the fruit. Whole wheat bread dough and seven compounds identified from the odour of bread dough synergize attraction of *Carpophilus lugubris, Carpophilus dimidiatus,* and *Carpophilus hemipterus* to its male pheromone. Odours from bread dough also synergize attraction of *Carpophilus davidsoni* to male pheromone. Propyl acetate, a host plant odorant synergizes attraction of *Carpophilus obsoletus* to its male pheromone. Compounds identified from host fruit synergize attraction of males and females to male pheromone in *Carpophilus hemipterus.* Host coattractants that synergize attraction to male pheromone were identified by Lin and Phelan (1991) from whole wheat bread dough inoculated with baker's yeast and from banana fruit inoculated with yeast.

Females of the cabbage looper moth, *Trichoplusia ni* are attracted to the sex pheromone of the male, and such attraction is enhanced or synergized by cotton foliage or by a ``hexane" extract of cotton foliage.

FEMALE-PRODUCED PHEROMONES

Sex attraction responses of males to female pheromone may also be directly influenced by stimuli from host plants. Male corn earworm moths are trapped in greater number when pheromone traps are placed near silking corn plants. Also, more *H. zea* males are trapped with a combination of a synthetic female sex pheromone blend and either Z-3-hexenyl acetate or Z-3-hexen-1-ol, both known host plant odorants, compared to female sex pheromone used alone.

More tobacco budworm *(Heliothis virescens)* are captured in traps baited with a combination of synthetic female sex pheromone and Z-3-hexenyl acetate than with pheromone alone. Similarly, male codling moths were trapped in greater numbers when traps baited with female sex pheromone were also baited with a blend of five compounds known to be green leaf volatiles.

Bark and ambrosia beetles in the genera *Dendroctonus, Scolytus,* and *Trypodendron* utilize female produced sex pheromones that are significantly enhanced by, or entirely dependent on, the release of host volatiles elicit behavioural responses. Females of the southern pine beetle, *Dendroctonus frontalis,* produce the pheromone ``*frontalin*" that may be inactive when released alone, but acts synergistically when the host tree compound α-pinene to attract predominantly males. The western pine beetle, *D. brevicomis,* uses female produced and male-produced compounds that are synergistically attractive when deployed together (neither is active alone) and are further enhanced in their attractiveness by the presence of the host compound myrcene. Females of the smaller European elm bark beetle, *Scolytus multistriatus,* produce multistriatin and 4-methyl-3-heptanol that are synergized by the host compound α-cubebene.

HOST PLANTS AND REPRODUCTIVE STRATEGIES OF MALES

Males of phytophagous insects most likely release pheromones attractive to females from host plants to optimize sexual encounter rates and mating opportuinites. Males of many species focus their mate-searching in areas of female activity, including host plants where females may oviposit, simply to maximize encounter rates (Thornhill and Alcock, 1983). Pheromone-releasing males may also maximize encounters with females by signalling (pheromonal calling) in areas of likely female activity, such as host plants where females may be feeding and ovipoisiting. In those species where host odorants synergize or enhance female attraction to male pheromones, the male must be on or near a source of such odorants when signalling in order to optimize his attractiveness to females.

Males may respond to male pheromone and to combinations of male pheromone with host odours in order to locate feeding sites and signalling sites. Male pheromones generally are an indication of the location of a resource as well as an areas of possible female activity. Another possible strategy of males calling from host plants is the advertising of resources to females. If females are attracted to male pheromone (with or without plant odorants) to gain access to the plant, then males are gaining increasing contacts with potential mates by advertising the presence of such a resource.

Advantages for males in selecting such sites for pheromone release may be (a) the likelihood that females may respond to host plant signals and be in the vicinity, (b) synergistic or enhancing effects of host odours on female responses to male pheromone, and (c) the arrival of females that are ready to mate on host plants for oviposition.

HOST PLANTS AND REPRODUCTIVE STRATEGIES OF FEMALES

Host plant influences on female pheromone behaviour may be strategies by females to maximize reproduction. Females of some species may use cues from host plants suitable for oviposition to regulate reproductive functions. Also, females may respond to pheromone and host odorants to locate or gain access to feeding and oviposition sites.

Females may synchronize reproductive activities via host plants, in part through plant stimuli acting on sex pheromone behaviour. Females may postpone sexual attraction, courtship and mating until suitable hosts are available. Hormonal regulation of sex pheromone production in the female corn earworm is controlled in part by signals from the host plant. Female corn earworm moths delayed pheromone biosynthesis and pheromonal signalling in the absence of host plant material and released greater amounts of pheromone in the presence of corn silk, because egg laying in this species commences immediately following mating, delaying mating effectively delays oviposition. Females that are not in the vicinity of suitable oviposition sites are not stimulated to oviposit, otherwise the offspring would be placed in poor habitats. A similar type of regulation may occur in species of *Yponomeuta* moths, which delay female pheromonal calling until on a host plant.

Females may also increase pheromonal signalling or begin such signalling upon encountering host plants, e.g. female cabbage looper moths on host plants begin calling earlier in the scotophase and are more attractive to males throughout the scotophase, compared with females not on host plants. Such findings indicate that the stimulation of calling and elevated levels of pheromone release are in response to plant cues. A female that is reproductively mature and is not on a host plant is unable to take advantage of such a resource and may suffer reduced reproduction if oviposition is thereby delayed. Observed stimulation of pheromonal calling in cabbage looper moths by a host plant may indicate a readiness to obtain a mate and to begin oviposition under those circumstances.

Thornhill and Alcock (1983) have also suggested that mating at an oviposition site may provide females access to males with a demonstrated ability to locate host plants, thus ensuring such abilities and greater fitness in offspring.

Baker (1989) noted that many male lepidopteran pheromones are structurally identical or are similar to chemical commonly found in plants. This observation appears to hold for male pheromones of phytophagous insects generally, including courtship and attractant pheromones. A number of chemicals isolated and identified from the hairpencils and scent scales of male moths and thought to be aphrodisiac or courtship pheromones occur widely in plants. sometimes as floral odorants. Pheromones of many bark beetles are terpenoids and are structurally related to host compounds. For example, cis-verbenol, a pheromone of *I. paraconfusus*, is derived from the host compound myrcene (Birch, 1984).

The major component of the pheromone of the male cabbage looper that is attractive to females is d-linalool. In fact, d-linalool was extracted from coriander seeds for use in cabbage looper attraction studies. The male pheromone of *G. molesta* is a blend of three chemicals that occurs in plants and the principal component is found in apple volatiles. The pheromones of the male medfly that are attractive to female include compounds that are terpenoids and fruity esters, similar to many compounds that are found in plants.

Male pheromones of phytophagous insects may be plant-related because females would have pre-existing receptors for such compounds and males can re-emit the same plant compounds to which females are sensitive, assuming that both sexes ingest them. Baker (1989) and Phelan, 1992) have suggested that female sensory apparati and receptor sites, developed for plant chemicals important to host finding and host selection, are readily available as pheromone receptors and may have evolved in response to pheromone chemistry. Many sex pheromones may then be structurally related to known plant chemicals because of a plant origin of prepheromone chemicals. It is well established that certain insect species use plant chemical directly as pheromones or as immediate precursors to pheromones. For certain species also male pheromones that advertise plants as feeding or oviposition sites for females may merely be according to host signals used by females as recognition cues, and they may thus be a type of mimicry, i.e. males on a host may mimic or enhance the odour of host plants to increase the probability of encounters with host-seeking females.

CONCLUDING REMARKS

Thus, it appears that the sexual communication in a diverse orders of phytophagous insects is strongly influenced by host plant chemistry. Well -documented examples are found in several families of the Coleoptera, Lepidoptera, and Diptera. The effects of host plants on sex pheromone behaviour are varied and reflect different strategies by males and females of different species. The majority of cases to date, however, involve male-produced pheromones that are attractive to both males and females and are released from the host plant or host material (Landolt and Phillips, 1997).

21

THE ROLE OF THE SENSE ORGANS IN MATING BEHAVIOUR OF COCKROACHES

It may be pointed out that insect mating behaviour consists of a series of behavioural interactions between the male and female individuals, finally leading to copulation. In general, male plays the more active role being both the initiator of activities during the courtship and showing a greater variety of them. The female often has a relatively passive role usually limited to behaviour patterns, signalling the acceptance or rejection of the male's advances. The various patterns of behaviour observed during courtship and copulation in cockroaches are released in most cases by chemicals and/or tactile stimuli. Mating behaviour has been studied in several cockroaches viz. *Byrsotria fumigata* (Barth, 1964), *Periplaneta americana, Blatta orientalis, Periplaneta australiasiae, Periplaneta burnnea* and *Periplaneta fuliginosa* (Barth, 1970).

The location of the sense organs involved in the receipt of stimuli releasing mating behaviour has been investigated in males of *P. americana* (Roth and Willis, 1952, Barth, 1961), *Blattella germanica* and *B. fumigata* (Barth, 1961). The role of sense organs in the mating behaviour of females have been studied on *Leucophaea maderae* (Engelmann, 1960, Smyth, 1963) and *P. americana* (Barth, 1961) and of both the sexes of *Nauphoeta cinerea, Leucophaea maderae, Byrsotria fumigata* and *Pycnoscelus surinamensis* (Roth and Barth, 1967).

In the following pages a brief account of the role of the sense organs in mating behaviour of cockroaches has been given according to Roth and Barth (1967).

SENSE ORGANS AND THE RESPONSIVENESS OF FEMALES

In *Nauphoeta cinerea, Leucophaea maderae* and *Byrsotria fumigata,* female sexual behaviour (mounting and feeding on the male's terga) is released by a male pheromone. The female's ability to respond rapidly to the courting male can be correlated with distribution of thin walled chemoreceptive sensilla on certain segments of the antenne. The antennal segments bearing chemoreceptive sensilla are recognizable under the low magnification by their rough appearance for example in females of *N. cinerea* chemoreceptors are abundant on antennal semgnets distal to No. 15 but not on those proximal to No. 15. It is, therefore, not surprising that females retaining to 10, 13 or 15 segments per antenna behave essentially as though they were completely antennectomized. The response of females to extracts of the male sex pheromone is also correlated with the distribution of chemoreceptive sensilla on the antennae.

In females of both *N. cinerea* and *L. maderae,* sensilla on the terminal segments of the maxillary and labial palps are capable of detecting the male. However, they are not as effective as the receptors on the antennae. Courting males were frequently observed to place ther abdominal tips directly beneath the mouth parts of antennectomized females. In these cases it is conceivable that mere tactile stimulation of the female's palpal receptors was sufficient to induce her to ``mount and feed on" his tergum but it seems likely that contact chemoreception is also involved. Engelmann (1960) found that antennectomized females of *L. maderae* failed to mate when kept with males for 26 days. His method of testing differed from Roth and Barth (1967) in that single females were kept with 2 males in finger bowls. But in the tests performed by Roth and Barth (1967) large numbers of both sexes were kept together in large cages. Under latter conditions, females would have a better opportunity of coming into contact with courting males. However, regardless of the number of males present, a female will not mate unless she responds to the courting male, ``feeds" on his terga, and allows him to grasp her genitalia (Engelmann, 1960; Roth and Barth, 1964).

In *B. fumigata,* the antennae appear to possess the sole receptors for the perception of stimuli from the male, since antennectomized females fail to mate. The behaviour of *P. surinamensis* is unique in that headless females are capable of mating. This curious situation is doubtless related to the fact that there is no preliminary courtship in this species. Females do not feed on the male's terga prior to copulation; they are almost completely passive, a situation that is most unusual in cockroaches.

SENSE ORGANS AND THE RESPONSIVENESS OF MALES

It has been pointed out that the antennectomized males of *P. americana* fail to respond to preparations of female sex pheromone (Roth and Willis, 1952; Barth, 1961). The same is true of antennectomized males of *B. fumigata,* though receptors on the palps apparently can detect the pheromone by means of contact chemoreception (Barth, 1961). Antennectomy in males of *B. germanica* reduces the percentage of males responding to sexual stimuli but does not affect the ability of responding individuals to discriminate between the sexes. The ability of these antennectomized males to discriminate between males and females persists in spite of maxillary and labial palpectomy (Roth and Willis, 1952). It should be noted that in addition to those located on the palps, contact chemoreceptors occur on the tips of the hypopharynx and ligula in *P. americana* and *B. germanica* (Frings and Frings, 1949), and these latter receptors may account, in part for the continued ability to discriminate between the sexes found in males of *B. germanica* lacking antennae and palps. In males of *Diploptera punctata,* bilateral antennectomy and palpectomy eliminated courting behaviour. Bilateral antennectomy alone reduced the percentage of males courting (within the one hour test period) and increased the average amount of time required (after contact with females) for the onset of courting behaviour. Bilateral antennectomy combined with unilateral palpectomy further increased the length of time required for the onset of courting behaviour but did not decrease the percentage of males that courted (Stay and Roth, 1956).

In males of *P. surinamensis,* the distance response to the female sex pheromone is correlated with the distribution of chemoreceptive sensilla on the antenna. These occur in large numbers from segent 9, distally. The weak response to the female pheromone shown by males retaining 9 segments per antenna is due to a small number of chemoreceptive sensilla present on these segments. However, as in *B. fumigata,* antennectomized males of *P. surinamensis* are capable of detecting the female pheromone through contact chemoreceptors located on the maxillary palps.

The behaviour of males of *N. cinerea*, L. *maderae*, and *B. fumigata* after ablation of various sense organs, differs from that of the females of these species in that it was impossible to eliminate mating by removing the obvious sense organs. The male's antennae bear the receptors which effect the normal rapid response to females. However, the maxillary and labial palps, and in some species perhaps the anals cerci as well, may play a role in the reception of stimuli from females but their importance can be demonstrated only by combined removal with the antennae, as the antennae are the organs of overriding importance. If these organs of secondary importance are removed in combination with antennectomy, the percentage of operated individuals mating is reduced and the length of time required before mating occurs is increased. The failure to eliminate mating in males of *N. cinerea and* L. *maderae*, possibly is related to the ease with which males of these species are induced to court by tactile stimuli alone, even in the absence of females. However, in *B. fumigata* release of courting behaviour by tactile stimuli alone is rare (Barth, 1964) and it is, therefore, probable that other receptors (on the hypoharynx and ligula) can detect the female sex pheromone after the antennae, palpi and cerci are removed (Roth and Barth, 1967).

THE ROLE OF THE ANTENNAE IN MATING BEHAVIOUR

It has been observed that antennal fencing between the male and the female is an important aspect of precopulatory behaviour in a variety of species of cockroaches (see : Barth, 1964). In addition to Engelmann's report (1960) that antennectomized females of *L. maderae* fail to mate, Smyth (1963) claimed that antennectomized *L. maderae* males never courted normal females, and that antennectomized females never responded to normal courting males. However, Roth and Barth (1967) indicated that antennal fencing is not essential in most cases since if given sufficient time, antennectomized females of *N. cinerea, L. maderae* and *P. surinamensis* do mate as do antennectomized males of all four species studied by Roth and Barth (1967). In *N. cinerea* mating enventually occurs even if both sexes are antennectomized.

It has been further observed by Roth and Barth (1967) that in *L. maderae* females attempting to ``feed" on the terga of anaesthetized males; similarly, normal males will court anaesthetized females that are completely inactive. Hence neither sex really requires active tactile stimulation (such as provided by mutual antennal fencing) for release of sexual behaviour. Roth and Willis (1952) found a similar situation in *B. germanica* females are normally active during courtship and mutual antennal fencing usually occurs. Stimuli resulting from these activities are undoubtedly important in arousing the male, yet some males do court immobile anaesthetized females. Thus, while the antennae are frequently not essential for successful mating, they do receive stimuli which aid in effecting rapid responses between courting individuals so that mating occurs without delay. In cockroach mating behaviour, as in many behaviour patterns, there appears to be a summation of external stimuli from various sources which, acting on the internal state of the organism most probably hormonally determined, serve to release the appropriate behaviour. Reduction of the sensory input e.g. by removing the antennae may increase the length of time required of reaching a level for excitation sufficient for release of a particular behaviour pattern. However, this reduction of sensory input does not prevent the eventual appearance of the behaviour in question owing to existence of other sources of input, which presumably acting together with fluctuations in the organism's internal state can produce a level of excitation sufficient for release of the particular behaviour pattern. In a sense the organism possesses a series of ``fail safe'' devices.

THE ROLE OF THE MALE'S WING RAISING DISPLAY IN MATING BEHAVIOUR

With the exception of *P. surinamensis*, the males of *N. cinerea, L. maderae* and *B. fumigata* exhibits the wing raising manoeuvre which is followed in the courtship sequence by the mounting and ``feeding'' activities of the female. These latter activities are released by a male sex pheromone. Release of the male pheromone is not contingent upon the raising of the wings for females of *N. cinerea* and *B. fumigata* attempt to mount courting males whose wings and tegmina have been cemented to the abdomen. Receptive females of *L. maderae* have been observed pushing their heads under or between the wings of motionless anaesthetized males in an attempt to gain access to his abdominal terga. Gluing the male's wings to his abdomen does, of course, prevent the female from achieving the correct orientation for mating. Courtship sequences involving such males rarely lead to successful mating even though both partners are receptive.

THE ROLE OF THE FEMALE'S MOUNTING AND FEEDING ACTIVITIES IN MATING BEHAVIOUR

The females mounting and feeding activities are important in stimulating the release of the male's grasping response so that actual copulation may be initiated (Roth and Willis, 1952, 1954; Barth, 1961, 1964). Males of *B. germanica* protrude their phallomeres when the isolated head of a female (mounted on a glass rod) is moved along their terga (Roth and Willis, 1952). In *B. fumigata,* phallomere, protrusion can be released by stroking the male's abdominal dorsum with brush while he is courting (Barth, 1964). Roth and Barth (1967) found that sealing the female's mouth parts with cement has little effect on mating in *N. cinerea* and *L. maderae.* In *N. cinerea*, mating sometimes occurs even when the male's wings and tegmina are cemented to his abdomen thus effectively preventing the female from reaching the first abdominal tergum. In *N. cinerea*, the lateral margins of the terga are exposed when the male's wings are folded in their normal position over the abdomen. If the wings and tegmina are glued in this position, the receptive female moves her mouth parts over the exposed portions of the terga, and copulation can sometimes be achieved even though the female is not in the normal position.

Hence, it is not essential that the male receives tactile stimulation on his first on second abdominal tergum for the release of phallomere protrusion. By contrast, mating was prevented in *B. fumigata* by cementing the male's wings and tegmina to his abdomen. In *B. fumigata* the male's wings cover the abdominal terga completely in the macropterous form and cover all but the terminal segments in the brachypterous form. Barth (1964) showed that males of *B. fumigata* require tactile stimulation on the first or second abdominal tergum for release of phallomere protrusion. The female's mounting and feeding activities are of considerable importance in all cockroaches showing this pattern of mating behaviour in that they orient the female correctly for copulation. If these activities are prevented, successful mating is rare or absent.

THE POSSIBLE ROLE OF OTHER SENSE ORGANS IN MATING BEHAVIOUR

In *B. germanica* as well as *P. americana* and *B. fumigata,* vision plays no role in sex recognition and subsequent mating behaviour (Barth, 1961, Roth and Willis, 1952).

Barth (1961) found that neither the male nor the female of *P. americana* requires intact cerci for normal mating behaviour. In his experiments, the cerci alone were removed, leaving all other sense organs intact. In the experiments performed by Roth and Barth (1967), the cerci of males have been shown to be able to respond to stimuli from females in *N. cinerea* and *L. maderae* (but not in *B. fumigata*) provided that the antennae and palpi are removed. Removing the cerci alone had no effect on mating. Hence, how the cerci function in mating behaviour is not known. In *N. cinerea* and other species of cockroaches, the female begins to feed at the posterior tip of the male's abdomen during courtship. Possibly, the cerci are mechanically stimulated at this time.

THE EFFECT OF NERVE CORD TRANSECTION ON MATING BEHAVIOUR

The effect of nerve cord transection on the mating behaviour differs in the two sexes. In females of *N. cinerea* and *L. maderae* nerve cord transection results in hypersexual activity (i.e. repeated matings) presumably because the mechanical stimuli resulting from the insertion of the spermatophore into the bursa copulatrix, which normally inhibit further mating behaviour, fail to reach the brain (Roth, 1962, 1964). The failure of males to mate after nerve cord transection may be due to a failure to receive the necessary stimulation from the female's ``feeding'' activities, but this seems unlikely since females often mounted the operated males, sufficiently so that their mouth parts reached the first tergum, a point anterior to the nerve cord transection. It seems likely that phallomere protrusion, the activity released by the female's mounting and ``feeding'' is reflexly coordinated by the last abdominal ganglion, but it may require stimulation from the brain hormone. If so, it is clear that the stimulus would not be able to reach its target in the operated insects. This possibility receives additional support from the fact, that nerve cord transected males of *P. surinamensis,* a species in which males do not require stimulation from female ``feeding'' activities for mating, mount the females in the characteristic fashion but do not protrude their phallomeres.

Roth and Barth (1967) have summarized the work on cockroaches as follows:

In *Nauphoeta cinerea, Leucophaea maderae* and *Byrsotria fumigata,* female sexual behaviour is released by a male sex pheromone. The antennae serve as distance receptors for the male odour. The female's ability to detect and respond to the male rapidly can be correlated with distribution of chemoreceptive sensilla of the antennae. The mounting and ``feeding" response of *N. cinerea* females to extracts of males was similarly correlated with the distribution of antennal chemoreceptors. In females of *N. cinerea* and *L. maderae* sensilla on the terminal segments of the maxillary and labial palps are capable of detecting the male in the absence of antennae. In *B. fumigata* females, the antennae alone appear capable of detecting the male. *Pycnoscelus surinamensis* is unique in that headless females can mate. This is explained by the fact that the females do not ``feed" on the male's terga prior to copulation.

It may be added that after ablation of various sense organs, the behaviour of males of *N. cinerea, L. maderae* and *B. fumigata,* differs from that of the females in that it is impossible to eliminate mating. The antennae bear the receptors which provide information promoting a rapid response to females. However, the maxillary and labial palps, and the cerci (except in *B. fumigata*) also play a role; the importance of these organs is shown only by combined removal with the antennae. The difficulty in eliminating mating in males of *N. cinerea* and *L. maderae* may be partly due to the fact that release of male courtship is not dependent upon stimuli from the female; these males may be induced to court by contact with other males. However, this is rare in *B. fumigata* and other sense organs (e.g. on the mouth parts other than the palps) may possibly be stimulated by the female sex pheromone. In *P. surinamensis,* the olfactory response to female sex pheromone is clearly correlated with distribution of chemoreceptive sensilla on the male's antennae. However, after bilateral antennectomy, the maxillary palps are capable of detecting the female on contact.

22

REPRODUCTIVE BEHAVIOUR

The reproductive behaviour involves the location of a mate, followed by courtship and mating, oviposition as well as parental or brood care.

LOCATION OF MATE

A wide variety of factors are involved in bringing together the sexes. These factors usually involve the visual, olfactory, auditory as well as tactile modes of communication singly or in combination.

The simplest visual pattern observed among water striders (Gerridae : Hemiptera) involves a male insect approaching any moving object of appropriate size that happens to enter its visual field. Other insects require more specific stimuli to induce approach e.g. male damselflies (Odonata) approach any insect that has transparent wings and flies in a manner similar to that of daselflies.

More complex visual stimuli involved in mate finding include the use of luminescent organs in signalling for a mate. This is best exemplified by the fireflies (Coleoptera Lampyridae) (Lloyd, 1971, 1977). In some species both the males and females produce light and have a very complicated signalling system, while in other species, only males produce the light signals. In either case, the light signalling systems are species specific. In fact, several species have been discovered through observation of differences in signalling systems. The mating behaviour of *Photinus pyralis* has been described in considerable detail by McElory and Selinger (1962). In some species of fireflies, several males congregate in a single tree or bush and flash synchronously (Buck and Buck, 1968). McElory *et al.,* (1974) have described this phenomenon in a remarkable way.

USE OF SEX PHEROMONES

The use of olfactory stimuli in mate location is widespread among the insect orders. Sex pheromones are produced by specialized glands in males, females or both sexes of a particular species. Sex pheromones or attractants are very potent substances that are used by the insect in a extremely small amounts. These attractants are undoubtedly among the most potent physiologically active substances known today (Jacobson, 1965). The sex pheromone, ``bombycol'', produced by the female, silkmoth, *Bombyx mori* can elicit a response from a male in a concentration of 1000 molecules per cubic centimetre. Electrophysiological studies of antennal receptors have revealed that the males are attracted from great distances and a single molecule of bombycol can initiate a single nerve impulse in a receptor cell (Schneider, 1974). Majority of the sex pheromomones are species specific. The production by females insects of pheromones that attract males has been observed and described for sevaeral species of insect orders (see : Jacobson, 1965, 1972, 1974).

Male bumble bees, *Bombus* species employ their mandibular gland-secretions in order to mark territorial sites which may play an important role in their mating behaviour. While the scent marks produced by male of one species do not attract males of other species and therefore, function as territorial markers, the marks do attract both males and females of the same species thus bringing the reproductives of both sexes together and increasing the probability that mating will occur. Most probably, the scent marks also physiologically probe the males and females for subsequent copulatory activity and thus play the role of sexual stimulations (Beroza, 1970).

It has been found that the virgin females of various species of silk moths (Saturniidae) call males by protruding posterior segments of the abdomen and exposing pheromone secreting glands to the atmosphere. This ``calling'' posture only occurs at certain times of the day or in response to certain stimuli and is controlled by the release of a calling hormone from neurosecretory cells of the corpora cardiaca (Riddiford and Truman, 1974). Female cockroaches (Dictyoptera) produce a sex pheromone that stimulates male alertness, antennal movement, searching locomotion and vigorous wing flutter (Jacobson, 1965).

As mentioned earlier, males in many species produce sex pheromones (see: Jacobson, 1965, 1974). Males of the *Galleria mellonella* secrete from glands on their wings an odorous substance that is very attractive to females. The bumble bee, *Bombus terrestris* males produce an attractant in their mandibular glands that attract females. The mandibular secretion has been identified as farnesol, a substance present in the flower oils of many plants (Jacobson, 1965).

Jacobson (1972, 1974) has described the occurrence of sex pheromones in female and male insects of the order Orthoptera, Hemiptera, Homoptera, Diptera, Isoptera, Neuroptera, Siphonaptera, Coleoptera, Hymenoptera, Lepidoptera; the anatomy and physiology of glands, the mechanism of attractant perception, responses to sex pheromones, influence of age on the production of and responses to sex pheromones; influence of time of day on sex pheromone production and mating etc. He has also given a comprehensive account of collection, isolation and identification of sex pheromones in various pterygote insects (viz, Orthoptera, Hemiptera, Lepidoptera, Coleoptera, Hymenoptera and Diptera). In addition, he has described the synthesis of sex pheromone and their use in insect control.

PHEROMONAL CONTROL OF MATING

Pheromones play a large part in both sexes during the mating of insects. They serve both as attractants from a distance and as aphrodisiacs exciting the opposite sex to copulation. Sometimes both effects are given by the same substances e.g. in the honey bee in which the queen substance ``9-oxodecenoic acid" serves also as a sex attractant which the drone will follow upwind and as an excitant which is necessary to provoke copulation. Sometimes the aphrodisiac secretion is most evident in the male. Many male Lepidoptera carry special scales (androconia) which are believed to produce an aphrodisiac secretion. The best evidence for this was obtained in *Eumenis semele* and other satyrid butterflies where the androconia of the male form an elongated patch across the centre of the forewing. These scent scales emit a scent which excites the female to accept the courting male. During an elaborate courtship, the knobs of the antennae of the female come to touch the patches of scent scales then mating immediately follows.

It has long been known that in cockroaches the dermal glands of both sexes give off pheromones which mix with the grease on the surface of the cuticle and act upon contact receptors on the antennae and mouth parts. In *Periplaneta* the sex attractant is produced mainly by virgin females, mating depresses its production and it decreases with age (Wigglesworth, 1970). In the females of

Byrsotria the corpus allatum control sex pheromone production. The pheromone releases courtship behaviour in the male, he raises his wings exposing the turgum, and apparently discharges a pheromone. The receptive female is attracted to the male and mounts and applies her mouth parts to his tergum.

Among Lepidoptera, there often seems to be a considerable overlap in the sexual scents of related species. It has been observed that a certain number of *Lymantria monacha* males were attracted to *Lymantria dispar* females in spite of strong competition from wild *monacha* females (Wigglesworth, 1970).

Secent organs in female Lepidoptera, occur close to the sexual opening and take the form of tufts of modified scales or hairs with gland cells at their bases or a simple fold in the body wall consisting of a glandular epithelium covered by thin cuticle devoid of pores. Such organs are best developed among those Lepidoptera, such as Saturniidae and Bombycidae in which the eggs of the females are ripe for laying at the time of emergence from the pupa. In such species the males are attracted over large areas to the newly emerged females. The antennae in the males are not only highly developed and plumose in structure but the sense organs they carry respond specifically to the sex attractant, which is not masked even in the presence of strongly smelling substances.

In the case of male mulberry silk moth *Bombyx mori* responding to ``bombycol'' the female sex attractant; 50% of the insects reacted by fluttering their wings in the presence of 10^{-4} μg of the pure substance and significant responses were given with 10^{-5} μg. A calculation of the lowest concentration of the substance in the air stream to which the antenne were exposed, that would give a response was 200 molecules per cm^3 when the behaviour threshold was used as the test; 10,000,000 molecules per cm^3 when the increase in nervous impulses, the electro antennogram threshold was used. At a concentration of 200 molecules per cm^3, 40 out of 40,000 specialized receptor cells will receive one ``bombycol'' strike per second (Schneider, 1974).

The major part of the active material is deposited on the gland surface ready for dissipation, because there is no large store. It is at a maximum at eclosion and production is suppressed after copulation. The behaviour of the female and the attractiveness of the gland secretion are not affected by allactectomy in the larval stage. The chemical structures of bombycol has been elucidated and it has been synthesized (see: Wigglesworth, 1970). It is trans-cis-10, 12-hexadecadienol.

The sex attractant obtained from the tip of the abdomen of the female gypsy moth, *Lymantria dispar* has been used for many years in surveys aimed at discovering the presence of the species in a given locality. This attractant named ``gyptol" has likewise been isolated and synthesized by Jacobson and Beroza (1963). It is similar to bombycol i.e. cis, 10-acetoxy, 7-hexadecenol.

The male wax moth, *Galleria mellonella* produces a sex attractant to which the female responds, it is secreted by a gland that lies in a fold at the base of the forewing. In this case, the active principle is *n*-undecanal. This belongs to the class of substances characteristic of repellent secretions in various insects and illustrates the specialized use of a common ingradient. This same phenomenon is perhaps to be seen also in the adult of the wireworm, *Limonius* where the male beetles are attracted from a distance of 12 m by the common fatty acid, valeric acid, produced by the female (Jacobson *et al.,* 1968), and in the mealworm *Tenebrio* where female sex pheromone which excites and attracts the male is in fact produced also by the male but in smaller quantities. In ants the different sexes and castes produce different odorous compounds, mixtures of terpenoids and indole bases. The mixtures are often qualitatively similar but each species or caste produces a blend of distinctive proportions.

Schowalter (2011) in his review has remarked that the sex pheromones greatly improve the efficiency with which insects find potential mates over long distances in heterogeneous environments. Pheromones are typically complex blends of compounds that may or may not be attractive to potential mates when present in ratios that are different from that produced by the "calling" sex. The particular blend of compounds and their enantiomers, as well as the time of calling, varies considerably among species. These mechanisms represent the first step in maintaining reproductive isolation. For example, among tortricids in eastern North America, *Archips mortuanus* uses a 90:10 blend of (Z)-11- and (E)-11-tetradecenyl acetate, *Archips argyrospilus* uses a 60:40 blend and *Archips cervasivoranus* uses a 30:70 blend. A related species, *Argyrotaenia velutinana* also uses a 90:10 blend but is repelled by the (Z)-9-tetradecenyl acetate that is incorporated by *A. mortuanus* (Carde and Baker, 1984). Among three species of saturniids in South Carolina, *Callosamia promethea* calls about 10:00-16:00, *C. securifera* about 16:00-19:00, and *C. angulifera* about 19:00-24:00 (Carde and Baker, 1984). Bark beetle pheromones have also been studied extensively (see Schowalter, 2011).

The evolution of male response to the pheromone blends which are emitted by females apparently is constrained by a trade-off between breadth of response and sensitivity (Hemmann *et al.*, 2008). Males which respond over a wide variation in pheromone composition likely fail to encounter conspecific females. Hence, selection should favour males with narrow response range, but high sensitivity. Hemmann *et al.* (2008) demonstrated, using a mutant pheromone strain of *Trichoplusia ni,* that hybrid males had the narrow breadth of wild males, but the low sensitivity of mutant males, suggesting a hybrid disadvantage and a mechanism for reinforcement of male pheromone response traits. Groot *et al.* (2006) reported that ten times more male *Heliothis virescens* were attracted to female *Heliothis subflexa* with introgressed quantitative trait locus from *Heliothis virescens* (that decreased the amount of acetate esters in pheromone glands) than to normal *H. subflexa* females. Hybrid infertility resulting from *H. virescens/H. subflexa* mating drove a strong directional selection for higher acetate ester concentrations in female *H. subflexa* pheromone.

Sex pheromones may be released passively, as in the faeces of bark beetles or actively through extrusion of scent glands and calling behaviour (Carde and Baker, 1984).

Pulses in emission and reception may facilitate orientation, because the antennal receptors require intermittent stimulation to avoid saturation and sustain upwind flight. However, Carde (1996) noted that the heterogeneous nature of the pheromone plume may make direct upwind orientation difficult over long distances. Pockets of little or no odour may cause the attracted insect to lose the odour trail. Detection can be inhibited further by openings in the vegetation canopy that create warmer convection zones or "chimneys", which carry the pheromone through the canopy (Fares *et al.,* 1980). Attracted insects may increase their chances of finding the plume again by casting, i.e., sweeping back and forth in an arcing pattern until the plume is contacted again (Carde 1996). Given the small size of most insects and limited quantities of pheromones for release, mates must be able to respond to very low concentrations. Release of less than 1 μg sec^{-1} by female gypsy moth, *Lymantria dispar,* or silkworm, *Bombyx mori,* can attract males,which respond at molecular concentrations as low as 100 molecules ml^{-1} of air (Harborne, 1994). Nevertheless, the likelihood of attracted insects reaching a mate is small. Elkinton *et al.* (1987) reported that the proportion of male gypsy moths responding to a caged female declined from 89% at 20 m distance to 65% at 120 m.

Of those males that responded, the proportion arriving at the female's cage declined from 45% at 20 m to 8% at 120 m, and the average minimum time to reach the female increased from 1.7 min at 20 m to 8.9 min at 120 m. Therefore, the probability of successful attraction of mates is low, and exposure to predators or other mortality factors relatively high, over modest distances.

Visual signalling is exemplified by the fireflies (Coleoptera : Lampyridae), (Fu *et al.*, 2005; Lloyd, 1983). In this group of insects, different species distinguish each other by variation in the rhythm of flashing, and by the perceived "shape" of the flashes that are produced by distinctive movements while flashing. Other insects, including glowworms (Coleoptera : Phengodidae) and several midges, also attract mates by producing luminescent signals (see also chapter 17).

USE OF AUDITORY ORGANS

Auditory or hearing signals serve to bring the sexes together for a large number of insects e.g. many Orthoptera. Heteroptera, Homoptera and Diptera some Lepidoptera and Coleoptera. Calling songs may be quite elaborate and are species specific. In general a response to one of these songs depends in part on the state of internal readiness for courtship and copulation e.g. the flight sound of a mature female mosquito is very attractive to a male and evokes copulatory behaviour. In some insect groups, the acoustical counter part of the ``assembling scent" has been found e.g. *Cicadas* in the genus *Magicicada* sing an aggregating song in chorus that is responsible for assembling both males and females (Alexander and Moore, 1962).

Cricket Song

The courtship songs of crickets are a familiar sound during summer and rainy season nights in every part of the country. The cricket makes these sounds by rubbing its wings together, thus rubbing the file of one wing across the scraper of the other. A lot of work has been done on the neurobiology of cricket stridulation (see : Alexander, 1968; Bentley and Hoy, 1975). It has been reported that only two thoracic ganglia nearest to the brain are necessary for singing.

COURTSHIP AND MATING

Courtship behaviour functions to bring together two insects of different sexes of the same species under conditions where mating is likely to occur and to be successful. The first problem is that of simply locating a potential mate. A variety of factors are associated with the mating behaviour of insects. Timing is an important part of successful mating : both the male and female must be in appropriate physiological conditions for mating. This is ensured by the synchronization of cycles that results from the interactions of environmental stimuli and of the two insects themselves. Furthermore, the mating systems vary greatly from species to species.

The production of the signals and the reactions to them constitute courtship behaviour in insects. Sounds, scent or sight causing the aggregation of insects may bring together a single male and female or may result in the accumulation of large numbers of either or both sexes. Apart from signals emitted by the insects themselves, swarming behaviour, in which a group of insects remains flying over one spot, may result from a reaction to some feature of the environment. Such are the swarms of small flies, common at sunset, holding station a few feet above a rock or a cow pat or under the tip of a branch silhouetted against the sky.

This behaviour is similar to that shown by males where they are crowded together or where there is a shortage of females. Aggression between male crickets, using a special song as a `weapon', is well known.

On some insects, such as the housefly `*Musca,* mating follows immediately upon recognition but more frequently the otherwise receptive female will not accept the male immediately and some further courtship is required. It is suggested that this female coyness may be important in allowing further time for correct recognition and certainly, in some cases, a precise sequence of signals is exchanged between male and female. A more obvious significance of courtship is seen in species where the female is aggressive; courtship involves some form of appeasement by the male so that he is not attacked (Blaney, 1977).

Among the most spectacular visual courtship patterns are those of fireflies and glow warms (Fig. 63 E,F). As the males fly over fields or woods, they display species - characteristic flash patterns. Females respond to the signal of a male of their own species with a brief flash. The latencies and signals are species - characteristic and temperature - dependent. The male responds to the female's signal and moves closer and closer as the pair continues to exchange signals, until the male comes to rest and mates.

Fig. 63E. Fireflies

Fig. 63F. A female Glow worm `calls' to her prospective mate with light signals. Here the female is wingless and resembles the larva. The male is fully winged and has well developed compound eyes.

After locating the mate (*vide supra*) in insects, males usually orient towards female which finally leads to copulation. Courtship behaviour may be simple or extremely complex. In many insects the sexes quickly copulate after mate location without any preliminary courtship management. Copulation in many insects may last longer without any specialized behavioural patterns. Beetles, flies and butterflies are often seen walking or flying in a team; some tree hoppers remain coupled for days (Horn, 1978).

Many of the cues that serve in mate location act further as releasers of courtship and copulation e.g. flight sound of the female mosquito not only attract males but also releases copulatory behaviour. In several insect species the sex pheromone produced by one of the sexes also serves as an excitant (i.e. aphrodisiac), stimulating courship or attempts at copulation e.g. the sex pheromone produced by female *Periplaneta americana* not only attracts males but also induces a wing raising display (Fig. 64) and attempts to copulate.

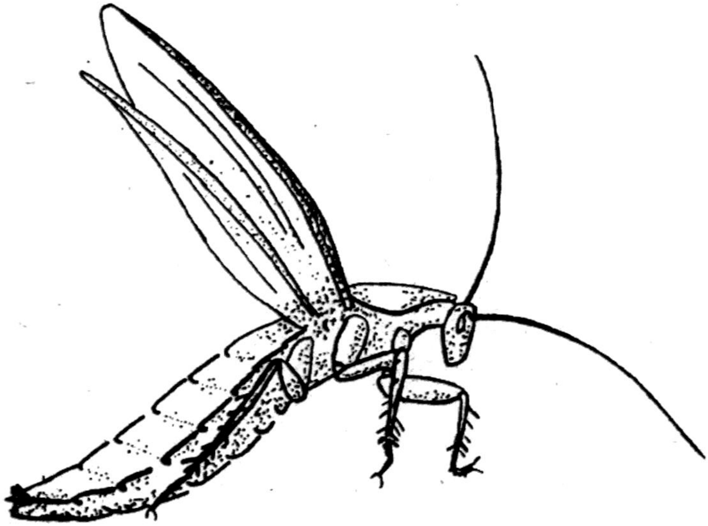

Fig. 64. Wing raising display of a male cockroach.

It has also been observed that the adult males of the cockroach, *Periplaneta americana* exhibit daily rhythms in responding to the sex attractant of virgin adult females. Peak responses were noted at 2.00 AM and the lowest responses were noted at 8.00 PM with normal photoperiod tests. In reversed photoperiod tests peak responses were at 2.00 PM and the lowest at 8.00 AM (see: Jacobson, 1972).

The sex pheromone is also capable of eliciting courtship behaviour in males of other species of *Periplaneta* and *Blatta orientalis*. In some aquatic insects, the sexually excited male or female individuals may produce a substance that acts as an excitant for the opposite sex e.g. the giant water bug, *Lethocerus indicus* (Jacobson, 1965). It has been remarked that the insects show an endless variety of sexual patterns. Some are simple i.e. the male and female simply coming together and copulating with little or no courtship managements. Jacobson (1965) has given a detailed account on this subject. Males of some species e.g. some *Drosophila* (Diptera), dragon flies (Odonata) and others congregate and display at a particular place. These aggregations of male individuals called `Leks' are presumably more effective in attracting female individuals rather than isolated male individuals.

INFLUENCE OF LIGHT ON MATING BEHAVIOUR

Experimental evidences have shown that the fruit fly, *Dacus tryoni* mates only at dusk. Before they mate, the flies behave in a characteristic way, both sexes fly actively and the males make a high pitched ``call''. When the flies were kept in cages at constant temperature but exposed to natural light, the characteristic premating behaviour began at dusk and continued for about an hour. Most of the flies took part in this activity and during this period some of them copulated, but at no other time did any of the flies become sexually active. Both the mating behaviour and copulation were confined to the period of dusk. It has also been observed that if during the half hour of dusk, when flies were sexually active, a light was turned on, they rapidly abandoned their activity and only to resume it again when light was switched off. Further, it has been observed that some species of *Drosophila* mate only in the light e.g. when adults of *Drosophila auraria, D. rufa* and *D. simulans* were kept continuously in the dark, none-mated. The beetles, *Aphodius howitti* mates only at dusk. Similarly the hepialid, *Oncopera fasciculata* is also crepuscular i.e. mate at dusk only. Mating follows a characteristic nuptial flight, which takes place for a brief period in the evening. Many species of mosquitoes have been observed and are known to form mating swarms during late afternoon or at dusk.

Different patterns of mating or copulatory behaviour of Dictyoptera, Heteroptera, Odonata, Lepidoptera (moths and butterflies), Orthoptera, Diptera and Hymenoptera are shown in Figs. 65 to 72.

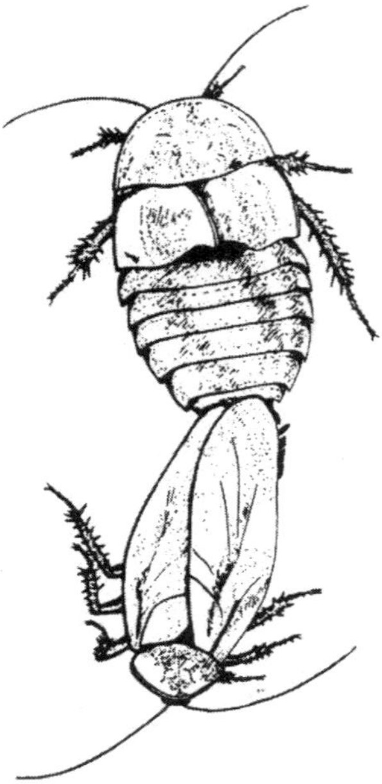

Fig. 65. Mating behaviour of cockroaches *Byrsotria fumigata* (Dictyoptera)

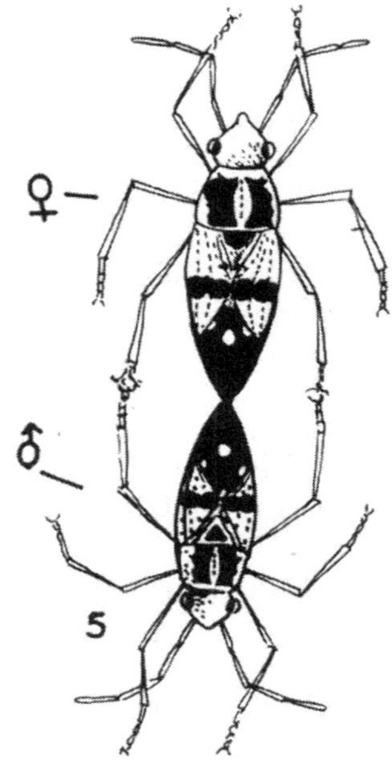

Fig. 66. Mating behaviour of milkweed bugs. *Lygaeus militaris* (Heteroptera)

Fig. 67. Mating behaviour of damselflies, *Calopteryx virgo* (Odonata)

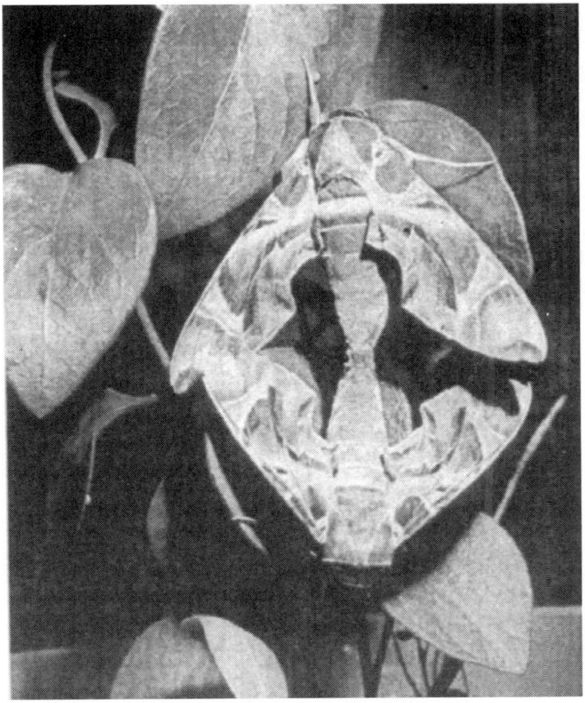

Fig. 68. Mating behaviour of Oleander Hawkmoths, *Daphnis nerii* (Lepidoptera)

Fig. 69. Mating behaviour of butterflies *Lasiommata* sp. The eye spots protect from predators (Lepidoptera).

Fig. 70. Desert locusts, *Schistoecrca gregaria* showing mating (A) and ovipositing (B) behaviour. The egg pod and eggs are also seen (C) (Orthoptera).

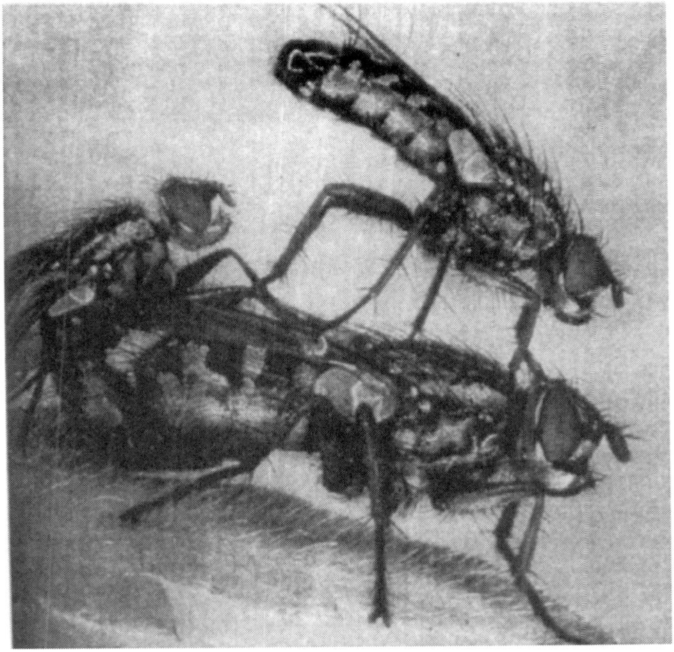

Fig. 71. Mating behaviour of fleshflies *Sarcophaga* sp. (Diptera)

Fig. 72. Mating behaviour of sawflies, *Cephus* sp. using its saw like ovipositor (Hymenoptera)

In the milkweed bug, *Lygaeus militaris* (Awasthi and Tiwari, 1975) mating takes place in hot summer days. A female bug is followed by 2 to 3 male bugs and all of them try to mount on the back of the female. The one which succeeds derives away all other bugs and they move underneath the fallen plant leaves. The mating takes place for several hours, sometimes more than 4 hours. During mating the metathoracic legs of male and the female lock together through their tarsii and beat the abdomen repeatedly (Fig. 66). It is believed that the beating of abdomen probably helps in the transfer of spermatophores from male to the female individual. After mating both the individuals separates. The female bug, which has once mated avoids further mating. Thus, excess of copulation is prevented. Probably mating occurs under the influence of either hormones or is governed by pheromones.

EGG LAYING BEHAVIOUR AS RELATED TO FOOD LOCATION

It has been commonly observed that though not always, the adult which is specialized for dispersal of the species, as well as it is in this stage of the life history that are found, the most remarkable adaptations for seeking out the right sorts of places to deposit eggs. For example, the butterflies, *Danaus archippus* (Danidae) and *Pieris rapae* (Pieridae) are remarkable for their widely ranging flights in search of suitable plants on which to oviposit. The larvae of *Danaus* live and feed on the milkweed, *Calotropis* species and are distributed widely in different parts of the country and more abundantly in the Rajasthan desert. But it is interesting to note here that during the summer months the larvae of *Danaus* are not seen on the *Calotropis* plants.

The *Lyctus* beetle-larvae, particularly *L. brunneus* lives in freshly seasoned timber, provided that it contains a large amount of starch and not less than 8% of moisture. Starch is absent from the true wood of all sorts of trees but it is present occasionally in adequate concentrations in the sap wood of certain broad leaved species. The *Lyctus* beetles search a place where the vessels of the xylem have been cut across or otherwise exposed and place their eggs inside the cavities of the vessels. Apart from rare accidents, the eggs are never laid anywhere else. The eggs of *Lyctus brunneus* are about 180 microns wide. The beetles usually select vessels about, 180 microns, they will not put their eggs in the wide vessels where they would be a loose fit, furthermore, they cannot put them in vessels that are narrower than the size of the eggs. Hence, the timber of coniferous trees and the true wood of broad leaved species do not have any eggs laid in them nor does the sap wood from the broad leaved trees in which the vessels of the xylem are too narrow. The first two would not contain enough starch to support the larvae but the last may in that case its unsuitability as a place for *Lyctus* to live depends entirely on the behaviour of the beeltes preventing from ovipositing any where except in vessels of such a size that the ovipositor can just be neatly pushed into them.

The mosquito, *Anopheles culifacies* lays its eggs in rice fields when the rice is not very high and when the rice plants reach a height of about 30 cms the mosquitoes disappear. The female mosquito while laying eggs flies a tortuous course just above the surface of the water. If it is prevented by the presence of tall rice plants from doing this, then the female mosquito will not lay eggs in that place. In connection with egg laying behaviour as related to food location, a mention may be made here of the selection of the suitable food.

It has been pointed out at several places that the larvae of the butterfly, *Danaus plexippus* feed only on milkweeds, *Calotropis* spp. The butterfly instinctively lays her eggs on the various parts of the milk weed leaves as well as inflorescence and virtually on no other sort of the plant. Since the eggs are placed so unerringly on the appropriate plant, one might not except to find in the larvae any special adaptation for recognizing its food. But they possess alfactory receptors on the antennae and maxillae. Unless these are stimulated by the smell of *Calotropis,* the caterpillars make no attempt to eat but will keep crawling until such time as they come upon a leaf of milkweed or die from starvation. When a caterpillar does come upon a leaf, it moves its head from side to side very close to the surface. If it receives the appropriate stimulus it tastes the leaf. The taste may then stimulate the caterpillar to continue feeding. (Andrewartha and Birch, 1954).

The blow fly, *Lucilia* lays its eggs on carrion at a specific (early) stage in its decompostion. With plenty of food, the maggots grow quickly, reaching a weight of about 60 mg before pupation. It is not essential for maggots to grow so big and when food is scarce, they may still pupate, provided that they can attain a weight of about 25 mg. When a few eggs are laid on a piece of meat of limited size, a few large flies are produced, with a moderate number of eggs, a larger number of smaller flies emerge. With increasing numbers of eggs laid on a fixed weight of meat, the number of flies emerging increasing up to a maximum and then begins to decrease eventually becoming zero.

Further, it has been found that when the adults of the ichneumonid, *Pimpla ruficollis* first emerge, their ovaries are small and they need to feed on the nectar from flowers of certain Umbelliferae before their eggs will develop to maturity and this takes about 3 weeks. The prey of *Pimpla* are the pine feeding caterpillars of *Rhyacionia buoliana*. While its ovaries were immature, *Pimpla* was repelled by pine oil and when the ovaries were mature *Pimpla* was attracted by pine oil. This beautiful adaptation led *Pimpla* away from the pine tree when it was seeking food for itself, and back to the pine when it was seeking food for its larvae and was ready to lay its eggs.

OVIPOSITION (EGG LAYING)

It is well known that a large number of insects are oviparous. The transfer of eggs from the ovary into the oviduct is followed by fertilization and then their deposition in the appropriate site. Sometimes there is no particular structure involved in the deposition of eggs but often the tip of the abdomen or appendages on it are modified to form a tube or ovipositor. The ovipositor may be rigid but more usually can be folded away or teliscoped when not in use. The ovipositor may be pointed or toothed so that eggs can be laid within the tissues of plant or animal hosts. Alternatively the eggs may be dropped passively, as in stick insects, or glued singly or in masses to some substrate. Some orthopterous insects deposit their eggs in packets or protective structures called oothecae. The protective case, secreted by the accessory glands, may be fronthy as in the locust or cuticle-like as in cockroaches. The process of ovulation is controlled by the hormones released from the brain. The oviposition i.e. deposition of eggs seems to be controlled by hormones as well as nervous system. The process of the egg passing from the external genital pore to the outside of the female is often associated with several behavioural processes like digging or probing into an egg laying location. The latter site is very important for the eggs because as soon as the larvae or nymphs as the case may be, emerge require immediate supply of the food or food plants of their choice depending on the nature of their feeding habits. So the suitable site for oviposition is a very important consideration for the female insect laying eggs. Occasionally the eggs are simply dropped to the ground or into water. In general eggs are deposited on or near to the food required by the offspring upon hatching.

In most insects, the adult is the active dispersal stage and females distribute eggs or larvae as the case may be usually in areas where the eggs or larvae will find favourable living conditions (Horn, 1978).

Horn (1978) has pointed out that male insects rarely take part in oviposition activity. An exception is Odonata in which males and females fly about one before the other. In dragon flies, the male leads and the female alternately removes spermatophores from the male abdomen and deposits eggs in the water. Dragon flies orient usually to reflected light from water and therefore, also attempt to oviposite on the roofs of shiny green automobiles. Sexual cooperation is even more vital in damsel flies. The female submerges to oviposit in under water plants and a joint effort by both sexes is needed to get her out of the water.

Thus, the survival, growth and development of immature insects depend to a large extent on oviposition in an appropriate environment. This is very important to those insects which require specific diets; e.g. female mosquitoes that are ready to oviposit must do so in a place where the eggs are in water. Although, some insects simply drop their eggs wherever they may happen to be e.g. Mayflies (Ephemeroptera). The beetle, *Hylotrupes* (Coleoptera : Cerambycidae) is attracted to terpene odour of the wood in which it deposits its eggs. The parasitic wasp *Nasonia,* locates the puparial cases of host flesh flies (Diptera : Sarcophagidae) by the odour of the decaying flesh in which the fly larvae and pupae are commonly found. *Pieris* butterflies (Lepidoptera : Pieridae) show a definite preference and occur in places where there are cruciferous plants for oviposition.

In some insects oviposition may be stimulated by exposure to light of a particular intensity or to darkness. It has been found in *Dacus tryoni* that 97% eggs were laid during periods when the lights were on. Isely and Ackerman (1923) found that the codling moth, *Cydia pomonella,* unlike *Dacus* laid most of its eggs during darkness, the moths which were kept in an insectary exposed to natural light laid about 93% of their eggs between sunset and dawn. When moths were kept continuously in darkness, eggs were laid more regularly throughout the day and night. When moths were exposed to alternating 2 hourly periods of light and darkness, 72% of the eggs were laid in the dark. Oviposition could be started at any time by putting the moths in the darkness and stopped at any time by putting them in the light. The cotton bollworms, *Earias* spp. and moths of the genus *Amsacta* also laid most of their eggs during darkness or twilight.

BROOD CARE

Once a parent insect has completed its responsibility for placing the eggs or larvae in an appropriate place, it may simply leave them without further notice i.e. in majority of the insects, parental care ceases with oviposition at a favourable spot (Figs. 73, 74, 75). However, some insects continue an association with the eggs and immature stages. This association is most highly developed in social insects e.g. ants, bees, wasps and termites, in which the brood form a core around which all activity is restricted and these brood are reared from egg to adult stage by workers.

Fig. 73. A beetle of family Buprestidae laying her eggs on the bark of a tree.

Fig. 74. A Sloe bug *Dolycoris baccarum* of Europe, with its batch of eggs stuck together and to the surface of a leaf. The eggs are laid on leaves of the Rose family, usually in batches of 28.

Fig. 75. A South African ladybird *Chilomenes lunulata* laying her eggs. A sticky secretion applied to each egg sticks them to the leaf surface. The eggs are laid on the leaves of plants frequented by aphids, the food of the larval ladybirds.

A small number of non-social insects continue association with their offspring, protecting or caring for eggs and immatures. In some giant water bugs (Heteroptera : Belostomatidae), the female oviposits on the dorsum of the male, which carries the eggs about until they hatch (Fig. 26 chapter 3). Awasthi (unpublished observations) found in the Ramgarh Lake of Gorakhpur, India, that in the water bugs like *Belostoma, Lithocerus,* and *Sphaerodema* the eggs are laid on the back of the male which carried them until they hatch. Presumably this has survival value in that a predator is less likely to take on father (male) water bug than the defenceless eggs (Horn, 1978).

Extended maternal care, including provision of food for offspring, is seen in crickets, cockroaches, some Hemiptera and Coleoptera, and non-social Hymenoptera. Maternal care can substantially increase growth and survival of offspring. Wood (1974) has shown that the female members of certain Membracidae (Homoptera) remains with her egg mass until it hatches and then she slices the bark of the host tree with her ovipositor, presumably to make it easier for nymphs to obtain plant juices. The nymphs stay in a cluster with their mother nearby (Fig. 27 chapter 3) and if a nymph wanders down the stem, mother membracid gently pushes it back to the group. She also defends the aggregation against predators (Horn, 1978). Survival of nymphs with their mother present was 80% compared to 60% when the mother was removed 2-3 days after egg hatch, and 10% when the mother was removed prior to making bark slits (slices). Females responded to predators or to alarm pheromones from injured offspring by fanning wings and buzzing, which typically drove the predator away (Wood, 1976).

A number of other non-social insects also do more for their offspring than merely deposit the eggs e.g. earwig females (Dermaptera) lay their eggs in burrows in the ground and guard them until they hatch (Fig. 28 Chapter 3). Some non-social insects e.g. solitary wasp captures an adult fly or any other insect and bring it to the nest (see : Evans, 1957). Following capture the wasp return to the nest and deposits an egg on the fly. Thereafter, a number of flies are brought to the nest as the larva grows. Finally, the larva spins a cocoon, where pupation takes place and ultimately emerges from the ground as an adult. Depending on species and the season it may remain in the coccon to tide over unfavourable conditions of the environment.

SELECTED BIBLIOGRAPHY

Ahearn, G.A. and Hardley, N.F. (1969). The effect of temperature and humidity on water loss in two desert Tenebrionid beetles. *Eleodes armola* and *Cryptoglossa verrueosa. Comp. Bioch. Physiol.* 30 : 739-750.

Alcock, J. 1975. *Animal Behaviour.* Sunderland, MA, Sinauer Associates.

Alenxander, R.D. 1964. The evolution of mating behaviour in Arthropods. In : *Insect Reproduction,* K.C. Highnam (ed.) pp. 78-94. *Roy, Ent. Soc.* London.

Alexander, R.D. 1967. Acoustical communication in Arthropods. *Ann. Rev. Entomol.* 12 : 495-596.

Alexander, R.D. 1968. Arthropods. In *: Animal communication, techniques of study and results of research* (T.A. Sebeok, ed.) Indiana University Press, Bloomington, Indiana USA, pp. 167-216.

Alexander, R.D. and Moore, T.E. 1962. The evolutionary relationships of 17 year and 13 year Cicadas and three new species (Homoptera : Cicadidae). *Univ. of Michigan Misc. Publ. Mus. Zool. No.* 121.

Allen, N., Kinard, W.S. and Jacobson, M. 1962. Procedure used to recover a sex attractant for the male tobacco hornworm. *J. Econ. Entomol.* 55: 347-351.

Alloway, T.M. 1972. Learning and memory in insects. *Ann Rev. Entomol.* 17 : 43-56.

Anonymous 2008. *Amazing book of facts.* Parragon Books, U.K..

Andrewartha, H.G. and Birch, L.C. 1954. *The distribution and abundance of animals.* The University of Chicago Press, Chicago and London.

Anonymous 2009. *The Book of Nature.* The natural heritage according to UNESCO pp. 428-429. Monaco Books. Verlag Wolfagang Kunt GMbH and Co., K.G.

Awasthi, V.B. 1982. Neurosecretory cells of the brain of Lepidoptera. *J. Adv. Zool.* 3 : 67-76.

Awasthi, V.B. 1985. Influence of juvenile hormone analogue on metamorphosis of *Philosamia ricini* Hutt. (Lepidoptera : Saturniidae). *Entomon* 10 : 165-169.

Awasthi, V.B. 2005. *Insect Neuroendrocrines.* Scientific Publishers (India), Jodhpur.

Awasthi, V.B. 2009. *Introduction to General and Applid Entomology, 3rd Edition,* Scientific Publishers (India), Jodhpur.

Awasthi, V.B. and Tiwari, Y.C. 1975. Studies on the mating behaviour of *Lygaeus militaris* Fabr. (Heteroptera : Lygaeidae). *Zool. Beitr.* 21 : 493-499.

Ayal, Y. 1994. Time lags in insect response to plant productivity : Significance for plant insect interactions in deserts. *Ecol. Entomol.* 19 : 207-214.

Baker, H.G. and Baker, I. 1973. Aminoacids in nectar and their evolutionary significance. *Nature,* 241 : 543 - 545.

Baker, H.G. and Baker, I. 1975. Studies of nectar constitution and pollinator plant coevolution. In : *Animal and plant coevolution,* pp. 100-140 (L.E. Gilbert and P.H. Raven, eds.). Austin University of Texas Press.

Baker, T.C. 1989. Origin of courtship and sex pheromones of the oriental fruit moth and a discussion of the role of phytochemicals in the evolution of Lepidopteran male scents. In: *Phytochemical ecology; Allelochemicals, Mycotoxins and insect pheromones and allomones.* (C.H. Chou and G.R. Waller, eds.) Taipi Acad. Seneca Monogr. Ser. No. 9.

Bartell, R.J. 1977. Behavioural responses of Lepidoptera to pheromones. In : *Chemical control of insect behaviour,* pp. 201-213 (H.H. Shorey and J.J. McKelvey Jr. Eds.) Wiley, New York.

Bartell, R.J., Shorey, H.H. and Barton-Brown, L. 1969. Pheromonal stimulation of the sexual activity of males of sheep blowlfly, *Lucilia cuprina* (Calliphoridae) by the female. *Animal Behaviour,* 17 : 576-583.

Barth, R. 1937. Bau und Function der Flugeldruseneiniger Mikrolepidopteren. Untersuchungen an den Pyraliden *Aphomia gularis, Galleria mellonella, Plodia interpunctella, Ephestia elutella, E. kuehniella, Z. Wiss. Zool.* 150 : 1.

Barth, R. 1938. Herkunft, Wirkung und Eigenschaften des Weiblichen Sexulduftstoffes einiger Pyraliden. *Zool. Jb. Physiol.* 58 : 297.

Barth, R.H. 1961. *Comparative and experimental studies on mating behaviour in cockroaches.* Ph.D. Thesis, Harvard University.

Barth, R.H. 1962. The endocrine control of mating behaviour in the cockroach, *Byrsotria fumigata, Gen. Comp. Endocrinol.* 2 : 53-69.

Barth, R.H. 1963. Endocrine-exocrine mediated behaviour in insects. *Proc. 16th Int. Congr. Zool.* 3 : 3-5.

Barth, R.H. 1964. The mating behaviour of *Byrsotria fumiata* (Guerin) Blattidae : Blaberinae), *Behaviour,* 23 : 1-30.

Barth, R.H. 1965. Insect mating behaviour : Endocrine control of a chemical communication system. *Science,* 149 : 882-883.

Barth, R.H. 1970. Pheromone-hormone, interaction in insects. *Mem. Soc. Endocrinol.* 18 : 373-404.

Barth, R.H. 1970. The mating behaviour of *Periplaneta americana* (Linnaeus) and *Blatta orientalis* Linnaeus (Blatteria : Blattidae) with notes on 3 additional species of *Periplaneta* and interspecific action of female sex pheromones. *Zeit. Tierpsychol.* 27 : 722-748.

Barth, R.H. and Lester, L.J. 1973. Neurohormonal control of sexual behaviour in insects. *Ann. Rev. Entomol.* 18 : 455-472.

Bates, H.W. 1862. Contribution to the insect fauna of Amezon valley. *Trans. Linn. Soc. London.* 23 : 495-566.

Beach, R. 1979. Mosquitoes biting behaviour inhibited by ecdysone. *Science* 205 : 829-831.

Beck, S.D. 1968. *Insect photoperiodism.* Academic Press, New York and London.

Beck, S.D., Edwards, C.A. and Mitler, J.T. 1958. Feeding and nutrition of the milkweed bug, *Oncopeltus fasciatus. Ann. Entomol. Soc. Amer.* 51 : 283-288.

Bennet-Clark, H.C. 1989. Songs and the physics of sound production. In: *Cricket behaviour and neurobiology* (F. Huber, T.E. Moore and W. Loher, eds.) Comstock Publishing Associates, A. division of Cornell Univ. Press, Ithaca and London, pp. 227 - 261.

Bentley, D.R. and Hoy, R. 1975. The neurobiology of cricket song. *Scientific American,* 231 : 34-44.

Bergstrom, G., Kullenberg, B., Stallberg-Stenhangen, S. and Stenhagen, E. 1968. Studies on natural odoriferous compounds. II. Identification of a 2, 3-dihydro farnesol as the main component of the marking perfume of male bumble bees of the species *Bombus terrestris. Arkiv. Kemi.* 28 : 453.

Beroza, M. 1970. *Chemicals controlling insect behaviour.* Academic Press, New York.

Billen, O. 1992. *Biology and evolution of social insects.* Leuven University Press, Leuven.

Birch, M.C. 1974. *Pheromones.* North Holland, Amsterdam.

Birch, M.C. 1984. *Aggregation in bark beetles.* In: *Chemical ecology of insects* (W.J. Bell and R.T. Crade, eds.). Sunderland M.A. Sinauer Associate.

Birukow, G. 1986. Orientation behaviour in insects and factors which influence it. In : *Insect behaviour* pp. 2-12 (P.T. Haskell, ed.). *Royal Entomol. Soc. London, Symp.* 3.

Blaney, W.M. 1977. How Insects Live. Galley Press.

Blest, A.D. 1957. The function of eye spot patterns in the Lepidoptera. *Behaviour.* 11 : 209-256.

Blest, A.D. 1977. The evolution of protective display in the Saturnoidea and Sphingidae (Lepidoptera). *Behaviour,* 11: 257-310.

Blomquist, G.J. and Vogt, R.G. 2003. *Insect Pheromones: Biochemistry and molecular Biology.* Elservier Academic Press, London.

Blum, M.S. 1966. Chemical releasers of social behaviour. VIII. Citral in the mandibular gland secretion of *Lestrimelitta limao. Ann. Entomol. Soc. Amer.* 59 : 962.

Blum, M.S. 1969. Alarm pheromones. *Ann. Rev. Entomol.* 14 : 57-80.

Blum, M.S. 1978. Biochemical defenses of insects. In: *Biochemistry of insects* (M. Rockstein, ed.) pp. 466-513. Academic Press New York.

Blum, M.S. and Brand, J.M. 1972. Social insect pheromones: their chemistry and function. *Amer. Zool.* 12 : 5553-5576.

Blum, M.S., Moser, J.C. and Cordero, A.D. 1964. Chemical releasers of social behaviour. II. Source and specificity of the odour-trail substance in four attine genera (Hymenoptera : Formicidae). *Psyche,* 71 : 1.

Blum, M.S., Padovani, F., Curley, A. and Hawk, R.E. 1969. Benzaldehyde : defensive secretion of a harvester ant. *Comp. Bioch. Physiol.* 29: 461.

Bohart, R.M. 1952. Pollination by native insects. In: *Year book of Agriculture,* USDA, Washington, DC. pp. 107-121.

Borsellino, A., Pierantori, R. and Schieti-Cavazza, B. 1970. Survival in adult meal worm beetles *(Tenebrio molitor)* of learning acquired at the larval stage. *Nature,* 225 : 963-964.

Brower, J.V.Z. 1958. Experimental studies of mimicry in some north American butterflies. The monarch, *Danaus plexippus* and the viceroy. *Limenitis archippus archippus. Evolution,* 12 : 32-47.

Brower, J.V.Z. 1958. Experimental studies of mimicry in some north American butterflies. Part III. *Danaus gilippus berenice* and *Limenitis archippus floridensis. Evolution,* 12 : 273-285.

Brower, J.V.Z. 1960. Experimental studies of mimicry IV. The reactions of starlings to different proportions of models and mimics. *Amer. Nat.* 94 : 271-282.

Brower, L.P. 1969. Ecological chemistry, *Scientific American*, 220 : 22-29.

Brower, L.P. and Brower, J.V.Z. 1962. Investigation into mimicry. *Nat. Hist.* 71 : 8-19.

Brower, L.P., Brower, J.V.Z. and Collins, C.T. 1963. Relative palatability and Mullerian mimicry among Neotropical Butterflies of the subfamily Helioconiinae. *Zooloigica,* 48 : 65-84.

Brower, L.P., Brower, J.V.Z. and Cranston, F.P. 1965. Courtship behaviour of the queen butterflies *Danaus gilippus berenice* (Craner). *Zoologica,* 50 : 1-39.

Brower, L.P., Ryerson, W.N., Coppinger, L.L. and Glazier, S.C. 1968. Ecological chemistry and palatability spectrum. *Science,* 161 : 1349-1351.

Brown, A.W.A. 1968. Insecticide resistance comes of age. *Bull. Entomol. Soc. Amer.* 14 : 3-9.

Broza, M. and Pener, M.P. 1972. The effect of corpora allata on mating behaviour and reproductive diapause in adult males of grasshopper, *Oedipoda miniata. Acrida* 1 : 79-96.

Buck, J.B. and Buck, E. 1968. Mechanism of rhythmic synchronous flashing of fireflies. *Science,* 159 : 1319-1328.

Buck, J.B. and Buck, E. 1976. Synchronous fireflies. *Scientific American,* 234 : 74-85.

Butler, C.G. and Simpson, J. 1967. Pheromones of the queen honey bee (*Apis mellifera*) which enable her workers to follow her when swarming. *Proc. Roy. Entomol. Soc. London.* 42 : 149.

Butler, C.G., Calam, D.H. and Callo, R.K. 1967. Attraction of *Apis millifera* drones by the odours of the queens of two other species of honey bees. *Nature,* 213 : 423.

Carde, R.T. 1996. Odour plumes and odour-mediated flight in insects. In: *Olfaction in Mosquito - host interactions. Ciba Foundation Symposium* 2009 pp. 54-70 John Wiley and Sons. Chichaster U.K.

Carde, R.T. 2011. *Insect Ecology: An Ecosystem Approach.* Third Edition. Elsevier - Academic Press.

Carde, R.T. and Baker, T.C. 1984. Sexual communication with pheromones. In: *Chemical ecology of Insects* (W.J. Bell and R.T. Carde, eds.) pp. 355-383. Chapman and Hall, London, U.K.

Carthy, J.D. 1958. *An introduction to the behaviour of invertebrates.* George Allen and Unwin Ltd. London.

Carthy, J.D. 1965. *The behaviour of Arthropods.* Freeman, San Francisco.

Casey, T.M. 1976. Flight energetics in phinx moth. Heat production and heat loss in *Hyles lineata* during free flight. *J. Exptl. Biol.,* 64 : 545-560.

Casey, T.M. 1981. Behavioural mechanisms of thermoregulation. In: *Insect thermoregulation* (B. Heinrich, ed.) John Wiley and Sons.

Cassier, P. 1965. Contribution al'etude des reactions photo menotaxique du criquet migrateur: *Locusta migratoria migratorioides* (R + F). leur determinisme sensoriel. *Insectes Soc.* 12: 363-382.

Cates, R.G. 1981. Host plant predictability and the feeding patterns of monophagous, oligophagous and polyphagous insect herbivores. *Oecologia,* 48 : 319-326.

Caussanel, C., Breuzet, M. and Karlinsky, A. 1978. Intervention du systame neuroendocrine cerebrale das le compartment parental de la famille de *Labidura riparia. C.R. Acad. Sci.* 286 : 1699.

Chapman, R.F. 1955. Some temperature responses of nymphs of *Locusta migratoria migratorioids* (R + F) with special reference to aggregation. *J. Exptl. Biol.* 32: 126-139.

Chinery, M. 2007. *Insects: Amazing images of fascinating creatures.* Octopus Publishing Group Ltd., London.

Clench, H.K. 1966. Behavioural thermoregulation in butterflies. *Ecology.* 47 : 1021-1034.

Cloudsley-Thompson, J.L. 1964. One of the functions of subelytral cavity in desert Tenebrionidae (Col.) *Entomol. Month. Mag.,* 60, 3-7.

Cloudsley-Thompson, J.L. 1964. On the Functions of the subelytral Cavity in desert. Tenebrionidae (Coleoptera). *Entoml. Mon. Mag.* 100:148-151

Cloudsley-Thompson, J.L. 1964. Terrestrial animals in dry heat. Arthropods. In : *Handbook of Physiology,* Sect. 4, pp. 447-499. (D.B. Dill, F.E. Adolph and C.G. Wilber, eds.) Americal Physiological Society, Washington, D.C.

Cloudsley-Thompson, J.L. 1965. *Desert Life.* Pergamon Press, Oxford.

Cloudsley-Thompson, J.L. 1967. *Microecology,* ELBS Edition.

Cloudsley-Thompson, J.L. 1975. Adaptations of Arthropods to arid environment. *Ann. Rev. Entomol.* 20 : 261-283.

Cloudsley-Thompson, J.L. 1977. Diurnal rhythms of locomotor activity in isolated desert locust, *Schistocerca gregaria* (Forsk.) *J. Inter-discip. Cycle Res.* 8 : 27-36.

Cloudsley-Thompson, J.L. 1991. *Ecophysiology of desert Arthropods and Reptiles.* Springer Verlag, Berlin.

Cloudsley-Thompson, J.L. 1993. *Diversity of desert life.* Scientific Publishers (India), Jodhpur.

Cloudsley-Thompson, J.L. and Chadwick, M.J. 1964. *Life in desert.* G.T. Foulis, London.

Conte, Y. Le and Hefetz, A. 2008. Primer Pheromones in social Hymeniptera. *Ann. Rev. Entomol.* 53: 523-542.

Coppel, H.C., Casida, J.E. and Dauterman, W.C. 1960. Evidence for a potent sex attractant in the introduced pine sawfly, *Diprion similis* (Hymenoptera). *Ann. Entomol. Soc. Amer.* 53 : 510.

Corbet, P.S. 1957. The life history of the emperor dragonfly *Anax imperature. J. Animal Ecol.* 26 : 1-69.

Corbet, P.S. 1960. Patterns of circadian rhythms in insects. *Cold spring Harbour Symp. Quant. Biol.* 25 : 357-360.

Corbet, P.S. 1963. A biology of Dragonflies. *Quadrangle.* Chicago.

Corbet, P.S. 1965. Asymmetry in cocrepuscular diel periodicities of insects. *Cand. Entomol.* 97 : 878-880.

Corbet, P.S. 1966. The role of rhythms in insect behaviour. In: *Insect behaviour* (P.T. Haskell, ed.) pp. 13-28, *Roy. Entomol. Soc. London.*

Corning, W.D., Dyal, J.A. and Willows, A.O.D. 1973. *Invertebrate learning. Vol. 2. Arthropods and Gastropod molluscs.* Plenum Press, New York

Cott, H.B. 1957. *Adaptive colouration in animals.* Methuen, London.

Crawford, C.S. 1981. *Biology of desert invertebrates.* Springer Verlag, Heidelberg, New York.

Crawford, C.S. 1988. Surface active Arthropods in a desert : Landscape. influence of microclimate vegetation and soil texture on assemblage structure. *Pedologia,* 32 : 373-385.

Crossley, S.A. 1974. Changes in mating behaviour produced by selection for ethological isolation between ebony and vestigial mutatants of *Drosophila melanogaster. Evolution,* 28 : 631-647.

Cullen, J.M. 1966. Ritualization of animal activities in relation to phylogeny, specialization and ecology. *Philos. Trans. Roy. Soc. London,* 251 (B) : 363-374.

Darwin, C. 1859. *The origin of species,* Menter edition 1963. Published by the New American Library of World Literature, Inc. New York.

Dean, J., Aneshansly, D. Edgerton, H.E. and Eismer, T. 1990. Defensive spray of the bombordier beetle : a biological pulse jet. *Science,* 248 : 1219-1221.

Deithier, V.G. 1966. Feeding behaviour. In : *Insect behaviour* (P.T. Haskel, ed.) *Royal Entomol. Soc. London,* Pp. 46-58.

Denlinger, D.L. 1986. Dormancy in tropical insects. *Ann. Rev. Entomol.* 31 : 239-264.

Dewsburry, D.A. 1978. *Comparative animal behaviour*. McGraw Hill Book Company, New York etc.

Digby, P.S.B. 1955. Factors affecting the temperature excess of insect in sunshine. *J. Exptl. Biol.* 32 : 279-298.

Dodson, C.H. 1975. Coevolution of orchids and bees pp. 91-99 In : *Animal and plant coevolution* (L.E. Gilbert and P.H. Raven, eds.) Univ. Texas Press, Austin, USA.

Draney, M.L. 1993. The subelytral cavity of desert tenebrionids. *Florida Entomologist,* 76 : 539-549.

Dupont- Raabe, M. 1956. Role des differents elements du systeme nerveus central dans la variation chromatigue des Phasmids. *C.R. Acad. Sci.* 243-1358.

Eckert, J.E. 1955. The flight of honey bee. *Amer. Bee. J.* 95 : 395-401.

Edney, E.B. 1937. A study of spontaneous locomotor activity in *Locusta migratoria migratorioides* (R & F) by the actograph method. *Bull. Entomol. Res.* 28 : 243-278.

Edney, E.B. 1971. The body temperature of Tenebrionid beetles in the Namib Desert of southern Africa. *J. Exptl. Biol.,* 55 : 253-273.

Edney, E.B. 1977. *Water balance in land Arthropods.* Springer, Berlin Heidelberg, New York, P. 282.

Edney, E.B., Haynes, S. and Gibo, D. 1974. Distribution and activity of desert cockroach. *Arenivaga investigata* (Polyphagidae) in relation to microclimate. *Ecology,* 55 : 420-427.

Ehrman, L. and Parsons, P.A. 1976. *The genetics of behaviour.* Sunderland M.A. Sinauer Associates, pp. 390.

Eidmann, H. 1956. Uber-rhythmische Erschinygen bei der Stabheuschrecke *Carausius morosus. Z. Vergl. Physiol.* 38 : 370.

Eisner, T. 1970. *Chemical defense against predation in Arthropods.* In: *Chemical ecology* (E. Sondheimer ad J.B. Simeone, eds.) Academic Press, New York.

Elkinton, J.S., C. Schal, T. Ono and R.T. Carde 1987. Pheromone puff trajectory and upwind flight of male gypsi, moths in a forest. *Physiol. Entomol.* 12: 399-406.

Elzinga, R.J. 1978. *Fundamentals of entomology.* Prentice Hall of India, New Delhi.

Engelmann, F. 1960. Hormonal control of mating behaviour in an insect. *Experientia,* 16 : 69-70.

Engelmann, F. 1970. *Physiology of Insect Reproduction.* Pergamon Press Oxford.

Estell, R.E. Fredrickson, E.L. and Havstad, K.M. 1996. Chemical composition of *Flourensia cerrua* at four growth stages. *Grass Forage Sci.* 51 : 434-441.

Evans, H.E. 1957. *Studies on the comparative ethology of the digger wasps of the genus Bembix.* Comstock Ithaca, New York.

Faegri, K. and Pijl, Van der, L. 1978. *The principles of pollination ecology.* 3rd edition, Pergamon Press, New York, pp. 242.

Fares, Y., P.J.H. Sharpe and C.E. Magnusen 1980. Pheromone dispersion in forests. *Jour. Theoret. Biol.* 84: 335-359.

Feeny, P.C. 1976. Plant appearance and chemical defense. *Recent Adv. Phytoch.* 10 : 1-40.

Fiori, G. 1977. La Cavita Sottoclitrale dei Tenebrionidi apmorti. *Redia,* 60 : 1-112.

Fletcher, B.S. 1977. Behavioural responses of Diptera to pheromones, allomones and kairomones. In : *Chemical control of insect behaviour,* pp. 129-148. (H.H. Shorey and J.J. McKelvey, Jr. eds.) Wiley, New York.

Flitters, N.F. 1964. The effect of photoperiod, light intensity and temperature on copulation, oviposition and fertility of Mexican fruitfly. *J. Econ. Entomol.* 57 : 811-813.

Foster, S.P. and Harris, M.O. 1997. Behavioural manipulation methods for insect pest management. *Ann. Rev. Entomol.* 42 : 123-146.

Fraenkel, G. 1929. Untersuehungen uber Lebensgewohnheiten, Sinnes Physiologie und Social psychologie der Wandernden Larven der Afrikanischer Wander - heusehrecke *Schistocerea gregaria* (Forsk.) *Biol. Zentralbl.* 49 : 657-680.

Fraenkel, G. and Gunn, D.L. 1940. *The orientation of animals: kineses, taxes and compass reactions.* Calender Press, Oxford.

Fraenkel, G. and Gunn, D.L. 1961. *The orientation of animals,* New York, Dover (reprint of 1940 Oxford edition).

Free, J.B. 1960. The behaviour of honey bees visiting the flowers of fruit trees. *J. Anim. Ecol.* 29 : 385-395.

Free, J.B. 1970. *Insect pollination of crops.* Academic Press, New York, pp. 544.

Frings, H. and Frings, M. 1949. The loci of contact chemoreceptors in insects : a review with new evidence. *Am. Middl. Nat.* 41 : 602-658.

Frisch, K.V. 1967. *The dance language and orientation of bees.* Harvard University Press, Cambridge, USA.

Frisch, K.V. 1971. *Bees-their vision, chemical senses and language,* Cornell University Press, Ithaca, New York.

Frisch, K.V. 1974. Decoding the language of the bee. *Science,* 185: 663-668.

Fuller, J.L. and Thompson, W.R. 1960. *Behaviour genetics.* Wiley, New York.

Furniss, M.M., Kline, L.N., Schmitz, R.F. and Rudinskey, J.A. 1972. Tests of three pheromones to induce or disrupt aggregation of douglass-fir beetles (Coleoptera: Scolytidae) on live trees. *Ann. Entomol. Soc. Amer.* 65 : 1227-1232.

Fu, X-H., N. Ohba, Y.Y. Wang and C.L. Lee 2005. Bioluminiscent courtship behaviour of aquatic firefly *Luciola substriata* (Gorh) Coleoptera Lamyridae) in China. *Acta Entomol. Sinica* 48: 227-231. Chinease with Engligh summary)

Gara, R.I. 1963. Studies on the flight behaviour of *Ips confusus* (Coleoptera : Scolytidae) in response to attractive material. *Contribu. Boyce. Thompson Instt.* 22 : 51.

Gary, N.E. 1976. Activities and behaviour of honey bees. In: *Hive and honey bees* (R.A. Grout, ed.) Dadant & Sons Hamilton, Illinois. USA.

Gatehouse, A.G. 1997. Behaviour and ecological genetics of wind borne migration by insects. *Ann. Rev. Entomol.* 42 : 475-502.

Gentry, C.R., Lawson, E.R. and Hoffman, J.D. 1964. A sex attractant in the tobacco budworm. *J. Econ. Entomol.* 57 : 819.

Gillett, J.D. and Haddow, A.J. 1957. Laboratory observations on the oviposition cycle in the mosquito, *Aedes africanus. Ann. Trop. Med. Parasitol.* 51 : 170-174.

Gillott, C. 1995. *Entomology.* 2nd ed., Plenum Press, New York and London.

Glick, P.A. 1939. The distribution of insects. spiders and mites in the air. *USDA Tech. Bull.* 67 : 1-150.

Goodden, R. 1971. *Butterflies.* Hamlyn.

Goodden, R. 1973. *All colour book of Butterflies.* Octopus Books Ltd., London.

Goodden, R. 1977. *The wonderful world of Butterflies and Moths.* Hamlyn.

Gould, J.L. 1975. Honey bee recruitment: The dance - language controversy. *Science* 189 : 685-693.

Gould, J.L. 1976. The dance language controversy. *Quart. Rev. Biol.* 51 : 211-244.

Groot, A.T., J.L. Horovitz, J. Hamilton, R.G. Santangelo, C. Sehal and F. Gould 2006. Experimental evidence for interspecific directional selection on moth pheromone communication. *Proc. Natl. Acad. Sci.* U.S.A. 103: 5858-5863.

Gullan, P.J. and Cranston, P.S. 1994. *The insects : An outline of entomology,* Chapman and Hall, London, etc.

Gullan, P.J. and Kosztarab, K. 1997. Adaptations in scale insects. *Ann. Rev. Entomol.* 42 : 23-50.

Gupta, J.K. and Gupta, P.R. 1997. Diversification of pollination. In: *Fruit crops pollination* (L.R. Verma, and K.K. Jindal, eds.) Kalyani Publishers, Ludhiana, New Delhi, etc.

Haddow, A.J. 1954. Studies on the biting habits of African mosquitoes. *An appraisal of methods employed with special reference to the twenty-four hour catch. Bull. Entomol. Res.* 45 : 199-242.

Haddow, A.J., Carbet, P.S. and Gillett, J.D. 1960. Laboratory observations on the oviposition cycle in the mosquito *Aedes apicoargenteus. Ann Trop. Med. Parasitol.* 54 : 392-396.

Haddow, A.J. and Gillett, J.D. 1957. Observations on the oviposition cycle of *Aedes aegypti. Ann. Trop. Med. Parasitol.* 51: 159-169.

Hadley, N.F. 1970. Micrometeorology and energy exchange in two desert arthropods. *Ecology,* 51 : 434-444.

Hadley, N.F. 1973. Desert species and adaptation. *Amer. Sci.* 60 : 1338-1347.

Hadley, N.F., Quintan, M.C. and Kennedy, M.L. 1991. Evaporative cooling in the desert Cicada. Thermal efficiency and water metabolic costs. *Jour. Exptl. Biol.* 159 : 269-283.

Halberg, F., Halberg, E. Barnum, C.P. and Bittner, J.J. 1959. Physiologic 24-hour periodicity in human beings and mice. The lighting regimen and daily routine. *Amer. Assoc. Adv. Sci.* 55 : 803-878.

Hamamura, Y. 1970. The substances that control the feeding behaviour and the growth of the silkworm, *Bombyx mori.* In: *Control of insect behaviour by natural products.* Academic Press, New York pp. 55-80.

Hamilton, W.J. 1973. *Life's color code.* McGraw Hill. New York.

Hamilton, W.J. 1975. Coloration and its thermal consequences for diurnal desert insects : In: *Environmental physiology of Desert organisms* (N.F. Hadley, ed.) 67-89. Dowden Hutchinson and Ross. Stroudsburg.

Hanson, F.E. 1970. Sensory responses of phytophagous Lepidoptera to chemical and tactile stimuli. In : *Control of insect behaviour by natural products.* Academic Press, New York, pp. 81-91.

Harborne, J.B. 1994. *Introduction to Ecological Biochemistry.* 4th Edition. Academic Press, London, U.K.

Harker, J.E. 1956. Factors controlling the diurnal rhythm of activity of *Periplaneta americana. J. Exptl. Biol.* 33 : 224-234.

Harker, J.E. 1961. Diurnal rhythms. *Ann. Rev. Entomol.* 6 : 131-146.

Harker, J.E. 1964. *The physiology of diurnal rhythms.* The university Press Cambridge.

Haydak, M.H. 1970. Honey bee nutrition. *Ann. Rev. Entomol.* 15: 143 - 156.

Heath, J.E. and Adams, P.A. 1965. Temperature regulation in the sphinx moth during flight. *Nature*, 205 : 309-310.

Heath, J.E. and Wilkin, P.J. 1970. Temperature responses of the desert cicada, *Diceroprocta apache* (Homoptera, Cicadidae) *Physiol Zool.* 43 : 145-154.

Heinrich, B. 1981. *Insect thermoregulation.* Jhon Wiley and Sons.

Hemmann, D.J., J.D. Allison and K.F. Haynes 2008. Trade-off between sensitivity and specificity in the cabbage looper response to sex pheromone. *Jour. Chem. Ecol.* 34: 1476-1486.

Henwood, K. 1975. A field tested thermoregulation model for two diurnal Namib desert tenebrionid beetles. *Ecology*, 56 : 1329-1342.

Hermann, H.R. and Blum, M.S. 1981. *Defensive mechanisms in the social* Hymenoptera. Academic Press, New York etc. pp. 77-197.

Hinton, H.E. 1960. Cryptobiosis in the larvae of *Polypedilum vnderplanki* Hint. (Chironomidae). *J. Insect. Physiol.* 5 : 286-300.

Hirsch, J. and Erlenmeyer-Kimling, L. 1961. Sign of taxis as a property of the genotype. *Science,* 134 : 835-836.

Hirsch, J. and Erlenmeyer-Kimling, L. 1962. Studies in experimental behavioural genetics. IV. Chromosome analysis for geotaxis. *J. Comp. Physiol. Psychol.* 55 : 732-739.

Hocking, B. and Sharplin, C.D. 1965. Flowers basking by arctic insects. *Nature,* 206: 215.

Holldobler, B. and Wilson, E.O. 1990. *The ants.* Springer Verlag, Berlin.

Hollis, J.H. 1963. Habituatory response decrement in pupae of *Tenebrio molitor. Animal Behaviour,* 11 ; 161-163.

Horn, D.J. 1978. *Biology of insects.* W.B. Saunders Co. London.

Howse, P.E. 1965. Nest building behaviour of the termites, *Zooteromopsis nevadensis* (Hagen). *Comp. Rend.* pp. 143-148.

Hoy, R.R. 1974. Genetic control of acoustic behaviour in crickets. *Amer. Zool.* 14 : 1067-1080.

Hoy, R.R. and Paul, R.C. 1972. Genetic control of song specificity in crickets. *Science*, 180 : 82-83.

Huber, F. 1962. Central nervous control of sound production in crickets and some speculations on its evolution. *Evolution,* 16 : 429-444.

Isely, D. and Ackerman, A.J. 1923. Life history of the codling moth in Arkansas with special reference to factors limiting abundance. *Bull. Arkansas Agr. Expt. Sta.* No. 189. p. 330.

Ishii, S. 1970. Aggregation of the german cockroach, *Blatella germanica* L. In: *Control of insect behaviour by natural products*, pp. 93-107, Academic Press, New York.

Jacobson, M. 1965. Insect orientation. *Ann. Rev. Entomol.* 8 : 94-114.

Jacobson, M. 1965. *Insect sex attractants.* Wiley Interscience, New York.

Jacobson, M. 1972. *Insect sex pheromones.* Academic Press, New York and London.

Jacobson, M. 1974. Insect pheromones, *In the physiology of insecta.* 2nd edn. 3 : 229-276. (M. Rockstein, ed.) Academic Press, New York.

Jacobson, M. and Beroza, M. 1963. Chemical insect attractants. *Science,* 140 : 1367-1373.

Jacobson, M., Toba, H.H., DeBolt, J. and Kishaba, A.N. 1968. Insect sex attractants VIII. Structure-activity relationships in sex attractant for male cabbage loopers. *J. Econ. Entomol.* 61 : 84.

Jander, R. 1963. Insect orientation. *Ann. Rev. Entomol.* 8 : 94-114.

Jaynes, J. 1969. The historical origins of ethology and comparative psychology. *Animal Behaviour*, 17 : 601-606.

Johnson, C.G. 1969. *Migration and dispersal of insects by flight,* pp. 763, Methuen, London.

Juberthie, C. and Caussanl, C. 1980. Release of brain neurosecretory products from the neurohaemal part of the aorta during egg-laying and egg-care in *Labidura riparia* (Insecta : Dermaptera). *J. Insect Physiol.* 26 : 427.

Kaiser, H. 1974. Die Regelung der individuendichte bei Libellen - mannchen (*Aeshna cyanea :* Odonata). Eine analyse mit systemtheorestichem. Ansatz. *Oecologia*, 14 : 53-74.

Karlson, P. and Butenandt, A. 1959. Pheromones (ectohormones) in insects. *Ann. Rev. Entomol.* 4 : 39-58.

Kennedy, J.S. 1961. A turning point in the study of insect migration. *Nature,* 189 : 785-791.

Kessel, E.L. 1955. The mating activities of balloon flies. *Systematic Zool.* 4 : 97-104.

Kevan, P.G. 1975. Sun-traking solar furnances in High Arctic Flowers: Significance for pollination and insects. *Science*, 189 : 723-726.

Kozin, R.B. 1976. *Pollination of entomophilous agricultural crops by bees.* Amerind Publ. Co. Pvt. Ltd. New Delhi and New York.

Krenn, H.W. 2010. Feeding mechanisms of adult Lepidoptera. Structure, function and evolution of mouth parts. *Ann. Rev. Entomol.* 155: 307-327.

Kugler, A. 1936. Die Ausnutzung der softmalsum farbung bie den Roszkastanienbluten durch Bienen and Hummeln. *Ber. Dtsch. Bot. Ges.* 60 : 128 - 134.

Kuhn, E.D. and Ambrose, J.T. 1982. Foraging behaviour of honey bees on golden delicious apple. *J. Amer. Soc. Hort. Sci.* 107 : 391-395.

Kumar, J. 1988. Insect pollinators in temperate fruits during bloom. *J. Tree. Sci.* 7 : 38-40.

Landolt, P.J. and Philips, T.W. 1997. Host plant influence on sex pheromone behaviour of phytophagous insects. *Ann. Rev. Entomol.* 42:371-391.

Law, J.H. and Regnier, F.E. 1971. Pheromones. *Ann. Rev. Biochem.* 40 : 533-548.

Levin, M.D. 1983. Value of bee pollination to U.S. agriculture. *Bull. Entomol. Soc. Amer.* 29 : 50-51.

Lin, H.C. and Phelan, P.L. 1991. Identification of food volatiles attractive to dusky sap beetle, *Carpophilus lugubris* (Coleoptera : Nitidulidae). *J. Chem. Ecol.* 17 : 1273-1286.

Lindauer, M. 1961. *Communication among social bees.* Harvard University Press, Cambridge, Mass. USA.

Lindauer, M. 1965. Social behaviour and mutual communication. In: M. Rockstein, ed.) *The Physiology of Insecta* 2 : 124-187. Academic Press, New York.

Lindauer, M. 1971. The functional significance of the honey bee waggle dance. *Am. Nat.* 105 : 89-96.

Lindauer, M. and Kerr, W.E. 1958. Die gegenseiting verstandingung beiden stachellosen Bienen. *Z. Vergl. Physiol.* 41 : 405.

Lindauer, M. and Kerr, W.E. 1960. Communication between the workers of stingless bees. *Bee World.* 41 : 29.

Lloyd, J.E. 1965. Aggressive mimicry in *Photuris* firefly femmes fatalis. *Science,* 149 : 653-654.

Lloyd, J.E. 1969. Flashes of *Photuris* fireflies, their value and use in recognizing species. *Florida Entomologist,* 52 : 29-35.

Lloyd, J.E. 1971. Bioluminescent communication in insects. *Ann. Rev. Entomol.* 16 : 97-122.

Lloyd, J.E. 1977. Bioluminescent and communication. In : *How animals communicate* pp. 164-183. (T.A. Sebeok, ed.) Indiana University Press, Blumington.

Lloyd, J.E. 1983. Bioluminescence and communication in Insects. *Ann. Rev. Entomol.* 28: 131-160.

Luscher, M. 1972. Environmental control of juvenile hormone (JH) secretion and caste differentiation in termites. *Gen. Comp. Endocrinol. Suppl.* 3 : 509-574.

Lutz, F.E. 1932. Experiments with Orthoptera concerning diurnal rhythms. *Amer. Museum Novitates,* 550 : 1-24.

Malcolm, S.B. 1990. Mimicry : status of a classical evolutionary paradigm. *Trends in ecology and evolution,* 5 : 57-62.

Manning, A. 1959. The sexual behaviour of two sibling *Drosophila* species. *Behaviour.* 15 : 123-145.

Mansingh, A. and Smallman, B.M. 1967. Effect of photoperiod on the incidence and physiology of diapause in two Saturniids. *J. Insect Physiol.* 13 : 1147-1162.

Markl, H. 1974. Insect behaviour: functions and mechanisms. In: *The physiology of insecta.* 3 : 3 - 148 (M. Rockstein, ed.) 2nd edn. Academic Press. New York.

Marsn, A.C. 1987. Thermal responses and temperature tolerance of a desert ant-lion larva. *Jour. Thermal. Biol.* 12 : 296-300.

Maschwitz, U.W. 1964. Gefahrenalarme stoffe und gefahrenalar-mierung bei Sozialen Hymenoptera. *Z. Vergl. Physiol.* 47 : 596.

Mathews, R.W. and J.R. Mathews 2010. *Insect Behaviour* 2nd Ed. Springer, Dordrecht, The Netherlands.

Matthews, E.G. 1976. *Insect ecology*. Univ. Queensland Press, St. Lucia, pp. 226.

May, M.L. 1976. Thermoregulation and adaptation to temperature in dragonflies. (Odonata : Anisoptera). *Ecol. Monogr.* 44 : 1-22.

May, M.L. 1978. Thermal adaptations of dragonflies. *Odonatologica.* 7 : 27-47.

McCleran, G.E. and DeFries, J.C. 1973. *Introduction to behaviour genetics.* Freeman, San Francisco.

McCluskey, E.S. 1958. Daily rhythms in male harvester and Argentine ants. *Science.* 128 : 536.

McCluskey, E.S. 1963. Rhythm and clocks in harvester and Argentine ants. *Physiol. Zool.* 36 : 273-292.

McCluskey, E.S. 1965. Circadian rhythm in male ants in five diverse species. *Science,* 150 : 1037-1039.

McConnell, J.V. 1966. Comparative physiology : Learning in invertebrates. *Ann. Rev. Physiol.* 28 : 107-136.

McElory, W.D. 1964. Insect bioluminescence. In : *The physiology of insecta* (M. Rockstein, ed.) 1 : 463-508. Academic Press, New York.

McElory, W.D. and Seliger, H.H. 1962. Biological luminescence. *Scientific American*, 207 : 76-89.

McElory, W.D., Seliger, H.H. and Deluca, M. 1974. Insect bioluminescence. In : *The physiology of insecta.* 2 : 411-460 (M. Rockstein, ed.) 2nd edn. Academic Press, New York.

McGregor, S.E. 1976. *Insect pollination of cultivated crop plants.* USDA-ARS-Agricultural Handbook, No. 496, pp. 411.

Menn, J.J. and Beroza, M. 1972. *Insect juvenile hormones : Chemistry and action.* Academic Press, New York.

Michel, R. and Bernard, A. 1973. Influence de la pars intercerebralis sur 1' induction an vol voutenu chez le criquet pelerin *Schistocerca gregaria. Acrida*, 2 : 139-149.

Michener, C.D. 1969. Comparative social behaviour of bees. *Ann. Rev. Entomol.* 14 : 299-342.

Michener, C.D. 1974. *The social behaviour of the bees : A comparative study.* Harvard Univ. Press, Cambridge, Massachusetts, pp. 404.

Milburn, N.S. and Roeder, K.D. 1962. Control of efferent activity in the cockroach terminal abdominal ganglion by extracts of the corpora cardiaca. *Gen. Comp. Endocrinol.* 2 : 70-76.

Milburn, N.S., Weiant, E.A. and Roeder, K.D. 1960. The release of efficient nerve activity in the roach, *Periplaneta americana* by extracts of corpus cardiacum. *Biol. Bull. (Woods Hole.)* 118 : 111-119.

Mishra, R.C. 1995. *Honeybees and their management in India.* ICAR, New Delhi.

Moore, B.P. 1968. Studies on the chemical composition and function of cephalic gland secretion in Australian termites. *J. Insect Physiol.* 14 : 32.

Moser, J.C. and Silverstein, R.M. 1967. Volatility of the trail - marking substance of the twon ant. *Nature,* 215 : 206.

Muller, J.F.T. 1879. *Ituna and Thyridia* : a remarkable case of mimicry in Butterflies. *Trans. Entomol. Soc.* (Proc.) 20-28.

Nielson, E.T. 1961. On the habits of the migratory butterfly. *Ascia monuste. Biol. Meddr.* 23 : 1-81.

Novak, V.J.A. 1975. *Insect hormones.* John Wiley and Co. New York.

Nunez, J.A. 1982. Honeybees foraging strategies at a food source in relation to the distance from the hive and rate of sugar flow. *J. Agric. Res.* 21 : 139-150.

Orshan, L. and Pener, M.P. 1979. Repeated reversal of the reproductive diapause by photoperiod and temperature in males of the grasshoppers. *Oedipoda miniata. Ent. Exp. Appl.* 25 : 219-226.

Orshan, L. and Pener, M.P. 1979. Termination and reinduction of reproductive diapause by photoperiod and temperature in males of the grasshoppers. *Oedipoda miniata. Physiol. Entomol.* 4 : 55-61.

Otte, D. and Joern, A. 1977. On feeding patterns in desert grasshoppers and the evolution of specilized diet. *Proc. Acad. Nat. Sci., Philadelphia,* 128 : 89-126.

Ouye, M.T. and Butt. B.A. 1962. A natural sex lure extracted from female pink bollworm. *J. Econ. Entomol.* 55 : 419.

Park, T. 1962. Beetles, competition and populations. *Science,* 138 : 1369-1379.

Pellmyr, O. 1992. Evolution of pollination and angiosperm diversification. *Trends in Ecology and Evolution,* 7 : 46-69.

Pener, M.P. 1965. On the influence of corpora allata on maturation and sexual behaviour of *Schistocerca gregaria. J. Zool.* 147 : 119-136.

Pener, M.P. 1967. Effect of allatectomy and sectioning of the nerve of the corpora allata on oocyte growth, male sexual behaviour and colour change in adults of *Schistocerca gregaria. J. Insect Physiol.* 13: 665-684.

Pener, M.P. and Broza, M. 1971. The effect of implanted active corpora allata on reproductive diapause in adult females of grasshopper, *Oedipoda miniata, Ent. Exp. Appl.* 14 : 190-220.

Phelan, P.L. 1992. Evolution of sex pheromones and the role of systematic tracking. In : *Insect chemical ecology. An Evolutionary approach* (B.D Roitberg and M.B. Ismann, eds.) Chapman and Hall, New York.

Proctor, M. and Yeo, P. 1972. *The pollination of flowers,* Taplinger, New York, pp. 418.

Pruthi, H.S. 1969. *A text book on Agricultural Entomology.* ICAR, New Delhi.

Raabe, M. 1982. *Insect neurohormones.* Plenum Press, New York.

Raabe, M. 1986. Insect reproductive regulation of successive steps. *Adv. Insect Physiol.* 19 : 29-154. Academic Press, New York.

Raabe, M. 1989. *Recent developments in insect neurohormones.* Plenum Press, New York.

Rabe, W. 1953. Beitrage zum orientierungsproblem der waserwanzerv. *Z. Vergl. Physiol.* 35 : 300-325.

Rajesus, R.S. and Reynolds, H.T. 1970. Demonstration of the presence of a female sex pheromone in the cotton leaf perforator. *J. Eco. Entomol.* 63 : 961.

Regnier, F.E. and Wilson, E.O. 1968. The alarm defence system of the ant, *Lasius alienus. J. Insect Physiol.* 15 : 893-896.

Rettenmeyer, C.W. 1970. Insect mimicry. *Ann. Rev. Entomol.* 15 : 43-74.

Rhoades, D. and Cates, R.G. 1976. Toward a general theory of plant antiherbivore chemistry. *Recent Adv. Phytochem.,* 10 : 168-213.

Ribbands, C.R. 1953. *The behaviour and social life of honeybees.* Bee Research Association Ltd., London, pp. 352.

Richards, O.W. 1951. The reaction to light and its inheritance in grain weevil, *Calandra granaria* (L.) Coleoptera : Curculionidae). *Proc. Zool. Soc.* London, 121 : 311-314.

Riddiford, L.M. and Truman, J.W. 1974. Hormones and insect behaviour. In *Experimental analysis of insect behaviour.* pp. 266-296. (L. Bartan Brown, ed.). Springer Verlag, New York.

Ritland, D.B. 1991. Unpalatability of Viceroy butterflies (*Limenitis archippus*) and their purported mimicry models, Florida Queens (*Danaus gilippus*). *Oecolgia* 88 : 102-108.

Robinson, G.E. 1992. Regulation of division of labour in insect societies. *Ann. Rev. Entomol.* 37 : 637-665.

Roeder, K.D. 1960. The release of efferent nerve activity in the roach, *Periplaneta americana* by extracts of the corpus cardiacum. *Biol. Bull.* (Woods Hole). 118 : 111-119.

Roeder, K.D. 1967. *Nerve cells and insect behaviour.* Harvard Univ. Press, Cambridge, Mass. USA.

Romoser, W.S. 1981. *The Science of Entomology.* 2nd edition, MacMillan Publshing, Co. Inc. New York.

Roth, L.M. 1962. Hyper sexual activity induced in female of the cockroach, *Nauphoeta cinerea. Science,* 138 : 1267-1269.

Roth, L.M. 1964. Control of reproduction in female cockroach with a special reference to *Nauphaeta cinerea.* I. First oviposition period. *J. Insect Physiol.* 10: 915-945.

Roth, L.M. and Barth, R.H. 1964. The Control of sexual receptivity in female cockroaches. *J. Insect Physiol.* 10 : 965-975.

Roth, L.M. and Barth, R.H. 1967. The sense organs employed by cockroaches in mating behaviour. *Behaviour,* 28 : 58-94.

Roth, L.M. and Willis, E.R. 1952. A study of cockroach behaviour. *Am. Middle. Nat.* 47 : 66-129.

Roth, L.M. and Willis, E.R. 1952. Possible hygroreceptors in *Aedes aegypti* (L.) and *Blatella germanica* (L.) *J. Morphol.* 91 : 1-14.

Roth, L.M. and Willis, E.R. 1954. The reproduction in cockroaches. *Smith. Misc. Coll.* 122 : 1-49.

Rothenbuhler, W.C. 1964. Behaviour genetics of nest cleaning in honeybees. IV. Response of F1 and backcross-generations to disease killed brood. *Amer. Zool.* 4 : 111-123.

Russell, A.S.H. 2007. *Whitaker's world of facts.* Penguine Books.

Sanders, C.J. 1971. Daily activity patterns and sex pheromone specificity as sexual isolating mechanisms in two species of *Choristoneura. Cand. Entomol.* 103 : 498.

Schmidt-Nelson, K. 1964. *Desert animals : Physiological problems of heat and water.* Clarenton Press, Oxford.

Schneider, D. 1974. The sex attractant receptor of moths. *Scientific American,* 231 : 28-35.

Schneirla, T.C. and Brown, R.Z. 1950. Army ant life and behaviour under dry season condition. IV. Further investigation of cyclic processes in behavioural and reproductive functions. *Bull. Amer. Mus. Nat. Hist.* 95 : 263.

Schowalter, T.D. 1996. Arthropod associates and herbivory on Tarbush in Southern New Mexico. *Southwest. Nat.* 41 : 140-144.

Schowalter, T.D. 2011. *Insect Ecology: An Ecosystem Approach.* Third Edition. Elsevier-Academic Press.

Seeley, T.D. 1983. Division of labour scouts and recruits in honey bee foraging. *Behav. Ecol. Sociobiol.* 12 : 253-259.

Seely, M.K. and Mitchell, D. 1987. Is the surface environment a thermal haven for chthonic beetles? *South Afr. J. Zool.,* 22 : 57-61.

Severin, H.H.P. 1933. Field observations on the Beet leafhopper, *Eufettix tenellus* in California. *Hilgardia,* 7 : 281-350.

Shorey, H.H. 1964. Sex pheromone of noctuid moths. II. Mating behaviour of *Trichoplusia ni* with special reference to the role of the sex pheromone. *Ann. Entomol. Soc. Amer.* 57 : 371-377.

Shorey, H.H. 1973. Behavioural responses to insect pheromones. *Ann. Rev. Entomol.* 18 : 349-380.

Slobodchikoff, C.N. and Wisman, K. 1981. A function of subelytral chamber of Tenebrionid beetles. *J. Exptl. Biol.* 90 : 109-114.

Smith, D.S. 1963. The organization and innervation of the luminescent organ in a firefly, *Photuris pennsylvanica.* (Coleoptera) *J. Cell. Biol.* 16 : 323-359.

Smith, W.E. and Whitford, W.G. 1976. Seasonal activity and Water loss relationships in four species of *Eleodes* (Coleoptera : Tenebrionidae). *Southwest Entomol.* 1 : 161-163.

Smyth, T.J. 1963. Mating behaviour of the Maderia cockroach. 35th Annual meeting of eastern Branch. *Entomol. Soc. Amer.* (Abstract).

Spieth, H.T. 1974. Courtship behaviour. *Ann. Rev. Entomol.* 19 : 383-405.

Stay, B. and Roth, L.M. 1956. The reproductive behaviour of *Diploptera punctata* (Blatteria : Diplopteridae) *Proc. Xth Int. Congr. Entomol.* 2 ; 547-552 (1958).

Steiner, K.E., Whitehead, V.B. and Johanson, S.D. 1994. Floral and pollinator divergence in the sexually deceptive sourth African Orchids. *Am. J. Bot.* 81 : 961-964.

Steiner, L.F., Mitchell, W.C., Harris, E.J. Kozuma, T.T. and Fujimoto, M.S. 1965. Oriental fruit fly eradication by mae annihilation. *J. Econ. Entomol.* 58 : 961-964.

Strang, F.E., Sheldahl, A., Hughes, P.R. and Hussein, E.M.K. 1970. Reproductive biology of *Lygus hesperus. Hilgardia,* 40 : 122.

Stuart, A.M. 1961. Mechanism of trail laying in two species of termites. *Nature,* 189 : 419.

Stuart, A.M. 1967. Alarm, defence and construction relationships in termites. *Science,* 156 : 1123.

Taber, S. 1964. Factors influencing the circadian flight rhythm of drone honey bees. *Ann. Entomol. Soc. Amer.* 57 : 769-775.

Takeda, M. and Skopik, D.S. 1997. Photoperiodic time measurement and related physiological mechanisms in insects and mites. *Ann. Rev. Entomol.* 42 : 323-350.

Tamaki, Y., Noguchi, H. and Yushima, T. 1969. Artificial control of mating activity of the smaller tea tortrix, *Adoxophyes orana.* Fischer von Roslerstamn and a quantitative bioassay for the sex pheromone. *Bochu-Kagaku,* 34 : 107.

Tewari, H.B. and Awasthi, V.B. 1968. The identical topography of the distribution of alkaline phosphatase and neurosecretory material in the neuroendocrine orgons of *Gryllodes sigillatus* (Walk.) Orthoptera : Gryllidae). *Gen. Comp. Endocrinol.,* 10: 330-338.

Thorpe, W.H. 1963. *Learning and instincts in animals.* 2nd edition. Methuen, London.

Thronhill, R. and Alcock, J. 1983. *The evolution of insect mating systems.* Harvard University Press, Cambridge, Mass.

Tigar, B.J. and Osborne, P.E. 1997. Patterns of Arthropod abundance and diversity in an Arabian desert. *Ecography,* 20 : 550-558.

Tigar, B.J. and Osborne, P.E. 1999. Patterns of biomass and diversity of aerial insects in Abu Dhabis sandy deserts. *J. Arid Environ.,* 43 : 159-170.

Tracy, C.R., B.J. Tracy and Dobkin, D.S. 1979. The role of pasturing in the behavioural thermoregulation by black dragons, (*Hagenius brevistylus* Selys : Odonata). *Physiological Zool.,* 52 : 565-571.

Traynier, R.M.M. 1970. Habituation of the response to sex pheromone in two species of Lepidoptera with reference to a method of control. *Ent. Exp. Appl.* 13 : 179.

Traynier, R.M.M. 1970. Sexual behaviour of the mediterranean flour moth, *Anagasta kuehniella* : Some influence of age, photoperiod and light intensity. *Cand. Entomol.* 102 : 534.

Truman, J.W. 1973. How moths ``turn on'' a study of the action of hormones on nervous system. *Amer. Sci.* 61 : 700-706.

Truman, J.W. 1973. Physiology of insect ecdysis II. The assay and occurrence of eclosion hormone in the chinese oak, Silkmoth. *Antheraea pernyi. Biol. Bull.* (Woods Hole). 144 : 200-211.

Truman, J.W. 1973. Physiology of insect ecdysis. III. Relationship between the hormonal control of eclosion and of tanning in the tobacco horn worm, *Manduca sexta. J. Exptl. Biol.* 58 : 821-829.

Truman, J.W. and Riddiford, L.M. 1974. Hormonal mechanisms underlying insect behaviour. *Adv. Insect. Physiol.* 10 : 297-352.

Truman, J.W. and Riddiford, L.M. 1979. Invertebrate systems for the study of hormonal effects on behaviour. *Vitamins and Hormones,* 25 : 283-315. Academic Press, New York.

Tweedie, M. 1973. *All colour book of insects.* Octopus Books Ltd., London.

Uvarov, B. 1977. Grasshoppers and Locusts. *A Handbook of General Acridology.* Vol. 2 : London : Centre for Overseas Pest Research.

Verma, R. and Jindal, K.K. 1997. *Fruit crop pollination.* Kalyani Publishers, Ludhiana.

Vite, J.P., Gara, R.F. and Kliefoth, R.A. 1963. Collection and bioassay of a volatile fraction attractive to *Ips confusus* (Coleoptera : Scolytidae). *Contrib. Boyce. Thompson Instt.* 22 : 39.

Volkonsky, M.A. 1939. Sur La Photo-akinase des acridiens. *Arch. Inst. Pasteur Alger.* 17: 194-220.

Waldbauer, G. 2006. *A walk around the Pond. Insects in and over the water.* Harvard University Press.

Waloff, Z. 1963. Field studies on solitary and *transcience* desert locusts in the Red Sea area. *Anti Locust Bull.* 40: 1-93.

Watt, W.B. 1968. Adaptive significance of pigment polymorphism in *Colias* butterflies. I. Variation of melanin pigment in relation to thermo-regulation. *Evolution.* 22 : 437-458.

Way, M.J. 1963. Mutualism between ants and honey dew producing Homoptera. *Ann. Rev. Entomol.* 8 : 307-344.

Weaver, N. 1978. Chemical control of behaviour intraspecific. In : *Biochemistry of insects.* pp. 360-389. (M. Rockstein, ed.) Academic Press, New York.

Weaver, N. 1978. Chemical control of behaviour, inter-specific. In : *Biochemistry of insects.* pp. 392-418. (M. Rockstein, ed.) Academic Press, New York.

Weber, N.A. 1972. Gardening ants - The Attines. *Amer. Physiol. Soc. Philadelphia,* pp. 164.

Wenner, A.M. 1962. Sound production during the waggle dance of the honey bee. *Animal Behaviour,* 10 : 70 - 95.

Whitford, W. 2002. *Ecology of Desert Systems.* Academic-Elsevier.

Wicker, W. 1968. *Mimicry in plants and animals.* Weidenfeld and Nicolson, London.

Wigglesworth, V.B. 1965. *The principles of insect physiology.* ELBS edition.

Wigglesworth, V.B. 1970. *Insect Hormones.* Freeman, San Francisco.

Wigglesworth, V.B. 1972. *The principles of the Insect Physiology.* 7th edition. Chapman and Hall, London.

Wigglesworth, V.B. 1974. *Insect physiology.* 7th edition, Chapman and Hall, London.

Wilde, J. de 1962. Photoperiodism in insects and mites. *Ann. Rev. Entomol.* 7 : 1 - 26.

Williams, C.B. 1930. *The migration of butterflies.* Oliver and Boyd, Edinburgh, pp. 473.

Williams, C.B. 1957. Insect migration. *Ann. Rev. Entomol.* 2: 163-179.

Williams, C.B. 1958. *Insect migration.* Collins, London, pp. 235.

Williams, N.H. 1991. Crop pollination and sustainable agriculture. *Bee World,* 72 : 45-46.

Williams, N.H. and Dodson, C.H. 1972. Selective attraction of male euglossine bees in orchid- fragrances and its importance in long distance pollen flow. *Evolution,* 26 : 84-85.

Wilson, E.O. 1958. A chemical releaser of alarm and digging behaviour in the ant, *Pogonomyrmex badius. Psyche,* 65 : 41.

Wilson, E.O. 1959. Source and possible nature of the odour trail of the fire ant, *Solenopsis saevisima, Science,* 129: 643.

Wilson, E.O. 1965. Chemical communication in social insects. *Science,* 149: 1064-1071.

Wilson, E.O. 1970. Chemical communication within animal species. *Chemical Ecology*, pp. 135-155. Academic Press, New York.

Wilson, E.O. 1971. *The insect societies.* Harvard Univ. Press, Cambridge, Mass, USA. pp. 548.

Wilson, E.O. 1972. On the queerness of social evolution. *Bull. Ent. Soc. Amer.* 19 : 20-22.

Wilson, E.O. 1975. *Sociobiology: The new synthesis.* Harvard University Press, Cambridge, Mass, USA.

Wilson, E.O. and Pawan, M. 1959. Source and specificity of chemical releasers of social behaviour in dolichoderine ants. *Psyche.* 66 : 70.

Wood, T.K. 1974. Aggregating behaviour of *Umbronia crassicornis* (Membracidae). *Can. Entomol.* 106 : 169-173.

Wood, T.K. 1976. Alarm behaviour of brooding female *Umbonia crassicornis* (Homoptera : Neurocidao) *Annals Entomol. Soc. Amer.* 69: 340-344.

Zera, A.J. and Denno, R.F. 1997. Physiology and ecology of dispersal polymorphism in insects. *Ann. Rev. Entomol.* 42: 207 - 230.

Zim, H.S. and Cottam, C. 1956. *Insects.* A guide to familiar American Insects. Golden Press, New York.

GLOSSARY OF SELECTED TECHNICAL TERMS

Aculeate Belonging to the aculeate Hymenoptera - wasps (including all bees) in which the ovipositor is modified as a sting.

Adaptation A change in structure, function or form that produces better adjustment of an animal to its environment.

Akinesis A state of immobility, due to lack of any stimuli.

Alate Possessing wings.

Alkaloids Chemicals found in plants, many with important pharmacological actions.

Allomone A communication chemical that benefits the producer by the effect it invokes in the receiver.

Altitude Height, measured as distance along the extended earths radius above a given datum, such as average sea level.

Ametabolous Lacking metamorphosis i.e. with no change in body form during development to adult, with immature stages lacking only genitalia.

Anemophily Wind pollination.

Anoxic Lacking oxygen.

Antenna (*pl.* antennae) Paired, segmented, sensory appendages, lying usually anterodrosally, on the head; derived from the second head segment.

Anthophilous Flower loving.

Apocritan Belonging to the suborder of Hymenoptera (Apocrita) in which the first abdominal segment is fused to the thorax.

Apolysis The separation of the old from the new cuticle during moulting.

Aposematic Warning of unpalatability, particularly using colour.

Aposematism A communication system based on warning signals.

Apterous Wingless.

Attractants Various chemical substances used to lure the insects.

Azadirachtin (AZ) The chemically active principle of the neem tree.

Batesian mimicry A mimetic system in which a palatable species obtains protection from predation by resembling an unpalatable species.

Benthos The bottom sediments of aquatic habitats and/or the organisms that live there.

Biological Control (also **biocontrol**) The humas use of selected living organisms to control populations of pest species (plant or animal)

Bioluminescence The production by an organism of cold light, commonly involving the action of an enzyme (luciferase) on a substrate (luciferin).

Bivoltine Having two generations in one year.

Borer The maker of burrows in dead or living tissue.

Brachypterous Having shortened wings.

Broad spectrum insecticide An insecticide with a broad range of targets, usually acting generally on the insect nervous system.

Bursicon A neuropeptide hormone that controls hardening and darkening of the cuticle after ecdysis.

Camoulflage Crypsis, in which an organism is indistinguishable from its background.

Cantharophily Plant Pollination by beetles.

Cardiopeptide A neuropeptide hormone that stimulates the dorsal vessel ('heart') causing haemolymph movement.

Castes Morphologically distinctive groups of indivudals within a single species of social insect, usually differing in behaviour.

Circadian rhythms Repeated periodic behaviour with an interval of about 24 hours.

Comb In a social hymenopteran nest, a layer of regularly arranged cells.

Conspecific Individual of the same species.

Coprophage A feeder on dung or excrement.

Corbicula The pollen basket of bees.

Corpora allata *(sing.* corpus allatum) Paired endocrine glands associated with the stomodaeal ganglia behind the brain the source of juvenile hormone.

Corpora cardiaca (*sing.* corpus cardiacum) Paired glands lying close to the aorta and ganglia behind the brain, acting as stores and producers of neurohormones.

Cosmopolitan Distributed worldwide.

Crepuscular Active at low light intensities, dusk or dawn.

Crypsis Camouflage by resemblance to environmental features.

Cryptobiosis The state of a living organism during which there are no signs of life and metabolism virtually ceases.

Cryptie species Those which are genetically and sexually distinct but cannot be differentiated on the basis of observed characteristics.

Cursorial Running or adapted for running.

Deciduous Falling off, detaching (e.g. at maturity).

Dermal gland A unicellular epidermal gland that may secrete cements, wax etc. and probably pheromones.

Detritivore (*adj.* detritivorous) An eater of organic detritus of plant or animal origin.

Diapause Delayed developed controlled by particular environmental conditions.

Diapause hormone A hormone produced by neurosecretory cells in the suboesophageal ganglion that affects the timing of future development of eggs.

Dispersal Movement of an individual or population away from its birth site.

Diurnal Active during the day.

Dufour's gland In aculeate hymenopterans, a sac opening into the poison duct near the sting site of production of pheromones and/or poison components.

Diversity Multiformity, unlikeness, of various kinds.

Ecdysis (*adj.* ecdysial) The final stage of moulting, the process of casting off of the cuticle.

Ecdysone The steriod hormone secreted by the prothoracic gland that is converted to 20-hydroxyecdysone, which stimulates the epidermal moulting.

Ecdysteroid The general term for sterioids that induce moulting.

Ecdysterone An old term for 20-hydroxyecdysone, the major moult-inducing steroid.

Eclosion The release of the adult insect from the cuticle of the previous instar; sometimes used of hatching from the egg.

Eclosion hormone A neuropeptide with several functions associated with adult eclosion including increasing cuticle extensibility.

Ecology The scientific study of interrelations between living organisms and their environment.

Ectoparasite A parasite that lives externally on, and at the expense of, another, which it does not kill.

Ectoparasitoid A parasite that lives externally on, and at the expense of, another, which it kills.

Endemic Restricted to a particular geographic area.

Endogenous rhythms Clock or calendar-like activity patterns, commonly circadian, unaffected by external conditions.

Endoparasite A parasite that live internally at the expense of another, which it does not kill.

Endoparasitoid A parasite that lives internally at the expense of another, which it kills.

Entomopathogen A pathogen (disease-causing organism) that attacks insect particularly.

Entomophage (*adj*. entomophagous) An eater of insects.

Entomophily Pollination by insects.

Entomophobia Fear of insects.

Environmental manipulation Alteration of the environment, particularly to enhance natural populations of insect predators and parasitoids.

Ethology The science dealing with the comparative study of animal behaviour.

Eusocial Exhibiting co-operation in reproduction and division of labour, with overlap of generations.

Exogenous rhythms Activity patterns governed by variations in the external environment (e.g. light, temperature, etc.)

Forage To seek and gather food.

Frass Solid excreta of an insect, particularly a larva.

Fungivore (adj. fungivorous) A fungus-eater.

Gall An aberrant plant growth produced in response to the activities of another organism, often an insect.

Gregarious Forming aggregations.

Gyne A reproductive female hymenopteran, a queen.

Habitat The sum total of environmental conditions of a specific place that is occupied by an organism, by a population, or a community.

Habituation Reduction in the response to a stimulus with repeated exposure, through modification of the central nervous system.

Haplodiploidy (adj. haplodiploid) A sex determination system in which diploid eggs give rise to femals, **haploid** ones to males.

Hemimetaboly With development in which the body form gradually changes at each moult, with wing buds growing larger at each moult; incomplete metamorphosis.

Hermaphroditism Having individuals that possess both testes and ovaries.

Homology (adj. homologous) Morphological identity or similarity of a structure in two (or more) organisms due to common evolutionary origin.

Honeydew A watery fluid containing sugars eliminated from the some Homoptera.

Hormone A chemical messenger that regulates some activity at a distance from the endocrine organ that produced it.

Host An organism that harbours another, especially a parasite or parasitoid, either internally or externally.

Host discriminate To choose between different hosts.

Host preference To prefer one host over another.

Hyperparasite A parasite that live upon another parasite.

Hyperparasitoid A secondary parasitoid that develops upon another parasite or parasitoid.

Hyporheic Living in the substrate beneath the bed of a waterbody.

Innate behaviour Behaviour requiring no choice or learning.

Insectivore An insect eater.

Instar The growth stage between two successive moults.

Integrated pest management (IPM) integration of chemical means of insect control with other methods, notably, **biological control** and habitat manipulation.

Juvenile hormone (JH) A 16, 17 or 18 C-chain hormone released by the corpora allata into the haemolymph, involved in many aspects of insect physiology, including modification of the expression of a moult.

Juvenile hormone mimics (*also* **juvenoids**) Synthetic chemicals that mimic the effect of JH on development.

Kairomone A communication chemical that benefits the receiver and is disadvantageous to the producer.

Kinesis (Pl. kineses) Movement of an organism in response to a stimulus, usually restricted to response to stimulus intensity only.

King The male primary reproductive in termites.

Klinokinesis Movement resembling a `random walk', in which changes in direction are made when unfavourable stimuli are encountered, with frequency of turning dependent upon the stimulus intensity.

Klinotaxis A movement in a definite direction relative to a stimulus, either directly towards or away from the source.

Labial palp A 1-5 segmented appendage of the labium.

Larva An immature insect after emerging from the egg, often restricted to insects in which there is complete metamorphosis (holometaboly) but sometime used for any immature insect that differs strongly from its adult.

Leaf miner A feeder on the mesophyll layer between the upper and lower epidermis of a leaf.

Lentic Of standing water.

Lotic Of flowing water.

Maggot A legless larval insect, usually with a reduced head, frequently a fly.

Matrifilial Of eusocial Hymenoptera whose colonies consist of mothers and their daughters.

Meconium The first excreta of a newly emerged adult following the pupal stage.

Melittophily Pollination by bees.

Menotaxis (including **light compass response**) Movement that maintains a constant orientation relative to a light source.

Metamorphosis The relatively abrupt change in body form between the end of immature development and the onset of the imaginal (adult) phase.

Microphage A feeder on small particles, such as spores.

Migration Mass movement, usually in search of more appropriate conditions.

Mimic One of the three components of a mimicry system, the emitter of false signal(s) received by an observer; an individual, population or species that resembles a model usually another species or part thereof; see also **Batesian mimicry; Mullerian mimicry.**

Mimicry The resemblance of a mimic to a model, by which the mimic derives protection from predation provided to the model (e.g. by unpalatability).

Model One of the three components of a mimicry system, the emitter of signal(s) received by the observer; the organism resembled by a mimic, protected from predation, for example, by distastefulness.

Monophage (adj. monophagous) An eater of only one kind of food, used particularly of specialized **Phytophages**.

Monophyletic Evolutionarily derived from a single ancestor, recognized by the joint possession of shared derived feature(s).

Moulting The formation of new cuticle followed by **ecdysis**.

Multiparasitism Parasitization of a host by two or more parasites or parasitoids.

Multivoltine Having several generations in one year.

Mullerian mimicry A mimetic system in which two or more unpalatable species obtain protection from predation by resembling each other.

Mycophage (adj. mycophagous) An eater of fungi.

Myiasis Diseases or injury caused by feeding of larval flies on live flesh, of humans or other animals.

Myremecophily Plant pollination by ants.

Naiad An alternative name for the immature stages of aquatic hemimetabolous insects.

Nasute A soldier termite possessing a snout.

Necrophage (adj. necrophagous) An eater of dead and or decaying animals.

Neem *Azadirachta indica* (Meliaceae) the neem tree.

Neotenic In termites, supplementary reproductives, arrested in their development, which have potential to take on the reproductive role should the primary reproductive be lost.

Neoteny The retention of juvenile features into the adjult stage.

Neurohormone The largest class of insect hormones, also known as neurpeptides, being small proteins secreted within different parts of the nervous systems.

Neurosecretory cells Modified neurones found throughout the nervous system producing insect hormones except ecdysteroids and juvenile hormones.

Niche The position or pattern of an organism within its community or ecosystem.

Nocturnal Active at night.

Nymph An immature insect after emerging from the egg, usually restricted to insect in which there is incomplete metamorphosis.

Obligatory Compulsory, or exclusive, e.g. obligatory diapause is a resting stage that occurs in every individual of each generation of a univoltine insect.

Oligophage (adj. oligophagous) An eater of few kinds of food, e.g. several plant species within one genus or one family; used particularly of phytophages.

Orthokinesis A response to a stimulus in which the rate of response (such as speed of movement) is positively proportional to the intensity of the stimulus.

Osmeterium (pl. osmeteria) An eversible tubular pouch on the prothorax of some larval swallowtail butterflies (Leipidoptera : Papilionidae) used to disseminate volatile toxic, defensive compounds.

Ovary One of the paired gonads of female insects, each comprising several **ovarioles**.

Oviparous Reproduction in which eggs are laid.

Ovipositor The organ used for laying eggs.

Paedogenesis Reproduction in an immature stage.

Palp Finger-like, usually segmented appendage of the maxilla **(maxillary palp**) and labium **(labial pulp)**.

Parasite An organism that lives at the expense of another (host) which it does not usually kill.

Parasitism The relationship between parasitoid or parasite and host.

Parasitization The condition of being parasitized, by either a parasitoid or a parasite.

Parasitoid A parasite that kills its host.

Parthenogenesis Development from an unfertilized egg.

Phalaenophily Plant pollination by moths.

Pharate Within the cuticle of the previous stage, cloaked.

Pheromone A chemical used in communication between individuals of the same species, releasing a specific behaviour or development in the receiver, pheromones play roles in aggregation, alarm, courtship, queen recognition, sex, sex attraction, spacing, dispersion and trailmarking.

Photoperiod The duration of the light (and therefore also dark) part of the 24-hour daily cycle.

Phylogeny (adj. phylogenetic) Evolutionary history of a taxon.

Physogastry Having a swollen abdomen, as in mature queen termites, ants and bees.

Phytophage (adj. phytophagous) An eater of plants.

Phytophagy The eating of plants.

Poison glands Accessory glands that produce poison as in the stings of Hymenoptera.

Pollination The transfer of pollen from male to female flower parts.

Polyphage (adj. polyphagous) An eater of many kinds of foods, e.g. many plant species from a range of families; used particularly of phytophages.

Polyphyletic A group evolutionarily derived from more than one ancestor, recognized by the possession of one or more features evolved convergently; rejected in cladistics and evolutionary systematics.

Predation Preying on other organisms.

Predator An organism that eats more than one other organism during its life.

Prey A food item for a predator.

Proboscis A general term for elongate mounth parts.

Prothoracic gland The thoracic glands that secrete ecdysteroids.

Prothoracicotropic hormone A neuropeptide hormone secreted by the brain that controls aspects of moulting and metamorphosis via action on the **corpora cardiaca**

Psychophily Plant pollination by butterflies.

Pupa The inactive stage between larva and adult in holometabolous insects; also termed a **chrysalis** in butterflies.

Pupation Becoming a pupa.

Pyrethrin One of the insecticidal chemicals present in the plant pyrethrum.

Quasisocial Social behaviour in which individuals of the same generation co-operate and nest-share without division of labour.

Queen A female belonging to the reproductive caste in eusocial or semi-social insects called a **gyne** in social Hymenoptera.

Quiescence A slowing down of metabolism and development in response to adverse environmental conditions.

Releaser A particular stimulus whose signal stimulates a specific behaviour.

Respiration A metabolic process in which substrates (food) are oxidized using molecular oxygen; also used inappropriately to mean breathing, as through spiracles or gas exchange across thin cuticle.

Rheophilic Liking running water.

Round dance A communication dance of honey bees.

Saprophage (adj. saprophagous) An eater of decaying organisms.

Semi-aquatic Living in saturated soils, but not immersed in free water.

Semisocial Social behaviour in which individuals of the same generation co-operate and nest share with some division of reproductive labour.

Semiochemical Communication chemical.

Sibling Full brother or sister.

Sociality The condition of living in an organized community.

Soldier In social insects, an individual worker belonging to a subcaste involved in colony defence.

Solitary Non-colonial, occurring singly or in pairs.

Species (adj. specific) A group of all individuals that can interbreed, mating within the group (sharing a gene pool) and producing fertile offspring, usually similar in appearance and behaviour.

Spatial Relating to space.

Spermatophore An encapsulated package of spermatozoa.

Sphecophily Plant pollination by wasps.

Spine A multicellular unjointed cuticular extension, often thorn-like.

Stadium The period between moults, the instar duration or intermoult period.

Stimulus An agent which provokes active response or reaction in a living organism.

Subsocial A social system in which adults look after immature stages for a certain period.

Superparasitism The occurrence of more parasitoids within a host than can complete their development within the host.

Symbiosis A long-lasting close and dependent relationship between organisms of two different species.

Synergism The enhancement of the effects of two substances that is greater than the sum of their individual effects.

Synomone A communication chemical that benefits both the receiver and the producer.

Systematic insecticides An insecticide taken into the body of a host (plant or animal) that kills insects feeding on the host.

Taxis (pl. taxes) An orientated movement of an organism.

Taxon Any taxonomic unit (species, genus, family, phylum, etc.)

Telotaxis Orientation and movement directly towards a visual stimulus.

Temporal Pertaining to time.

Trophic Relating to food, of an egg of a social insect, degenerate and used in feeding other members of the colony.

Univoltine Having one generation in one year.

Variation The occurrence of heritable or non-heritable differences among the individuals of one population or among populations.

Voltinism The number of generations per year.

Waggle dance A communication dance of honey bees.

Wasmannian mimicry A form of mimicry which allows an insect of another species to be accepted into a social insect colony.

Wax A complex lipid mixture giving water proofing to the cuticle, providing covering or building material.

Weed Any plant `in the wrong place', particularly used of plants away from their natural range, or invading human monocultural crops.

Wood borer An insect that tunnels into live or dead wood.

Xylophage An eater of wood.

Zoophilic Preferring other animals to humans, especially used of feeding preference of blood-feeding insects.

Appendix - 1

AMAZING INSECTS

Most abundant : Insects called springtails live in topsoil all over the world. There are probably as many as 600 million per hectare. Together they weigh more than the entire human race.

There can be up to five billion aphids per hectare in a swarm.

A swarm of desert locusts may contain 50 billion insects. One seen in 1889 was reckoned to have 250 billion locusts weighing a total of half a million tonnes.

Fastest : Dragonflies can fly at speeds of 28.7 – 32 km/h.

Heaviest : Goliath beetles can be 11 cm long and weigh 100 g.

Longest Wingspan : The giant silkmoth *Attacus atlas* is found from India eastward through China to Malaysia. It is reputed to be the largest moth in the world. It attains a wing expanse of 12 inches. Likewise, the Queen Alexandra's birdwing of Papua New Guinea is the largest butterfly which can be 12 inches across.

Longest: The world's longest insect is the giant stick insect of Indonesia which grows upto 13 inches long.

Smallest : The world's smallest insect is *Mymar* 0.2 mm long (a bettle of Trichopterigidae) parasitic in eggs of insects.

Longest – Lived : Jewel beetle larvae, or young, may live inside timber for 30 years or more before becoming adult insects.

Shortest-Lived : Mayflies may live for a single day as adults, but for two or three years as larvae.

Noisiest : Colonies of hundreds of male cicadas make loud noises to attract females. These sounds are as noisy as heavy traffic.

Long-distance fliers : Butterflies have been tracked travelling 4,828 km.

(Goodden, 1977; Russell, 2007; Anonymous, 2008).

Appendix - 2

BEE FACTS

There are about 20,000 different types of bee. The best known is probably the honeybee, but not all bees live in colonies like the honeybee. Many live alone and build their own nests.

Wallace's giant bees are the world's largest at upto 4 cm long. They were first found in Indonesia in 1858 but were then thought to have become extinct. In 1981 the bees were rediscovered.

The smallest bee is the Brazillian *Trigona duckei* at 2-5 mm long.

Honeybees are the only insects that make food which humans eat. Bees make honey to feed the inhabitants of their hives during winter. The honey they do not eat is harvested by beekeepers. Bees also make beeswax which is used for making candles and furniture polish.

One beehive may contain 50,000 worker bees-these are the bees which collect nectar from flowers to make honey.

Honeybees visit upto five million flowers to make one kilo of honey. They fly a total distance equal to flying four times round the Earth.

Queen bees can lay two or three thousand eggs a day- as many as 200,000 eggs a year. They may live as long as five years, so can produce a million eggs.

Honeybees' wings beat 11,400 times a minute. These wing movements make the bees' familiar buzz.

1.3 million tonnes of honey are produced worldwide every year. China is the main producer with 276,000 tonnes, followed by the USA with 82,000.

Some people are allergic to bee stings, which can even be fatal. In 1962 in Rhodesia, Johanne Relleke was stung 2,243 times by wild bees, but survived. (Russell, 2007)

Appendix - 3

UNIVERSITY OF CAMBRIDGE · DEPARTMENT OF ZOOLOGY
DOWNING STREET · CAMBRIDGE CB2 3EJ
TELEPHONE 022-358717

VBW/JES

10th August, 1976

Dear Dr. Awasthi,

I entirely agree with what you set out in your letter about the definition of neurohumors etc. The exceedingly condensed little chapter which I added to the last edition of Insect Physiology was based on my earlier book, Insect Hormones. In the book I completely avoided the term neurohormones, and I would suggest you draw a line through the words "neurohormones" on page 155 of Insect Physiology.

On the whole, I think Berta Scharrer's definitions of 1969 are good. I myself, however, do not have a passion for definitions, which always lead you into difficulties where distinctions are still not clear. Scharrer herself admits the "border-line cases which more or less defy classification".

In my section on 'neurohumors' in Insect Hormones, I point out that both noradrenaline and acetyl choline, besides acting at neuromuscular junctions in vertebrates, are also active in the circulating blood, and indeed adrenaline and noradrenaline are among the classic examples of "hormones".

When it is fully established that all neurosecretory cells are in fact giving rise to peptide hormones, it may be convenient to group these altogether as neurohormones, but the function of very many neurosecretory cells are still quite unknown, and that was why I avoided the use of the word "neurohormones".

But I repeat that I am in full agreement with what you set out in your letter.

Yours sincerely,

V.B. Wigglesworth

Dr. V.B. Awasthi,
Department of Zoology,
University of Jodhpur,
INDIA.

P.S. I am looking out a few reprints to send you

SUBJECT INDEX

ACKNOWLEDGEMENT AND FIGURE SOURCES

A Denis 37A,B,C, 41B

C.S. Crawford 56A

D.J. Horn 27, 60A

G. Waldbauer 26

H.S. Zim and C. Cottam 4, 7A,B, 16A,B, 17, 19-A-D, 22, 25A,B,C, 40, 43, 44A,B, 47, 56 B, C, 57 C, D, 63E.

H.S. Pruthi 23A,B,C

J. Brackenbury 35, 36A,B, 42, 53, 54A,B,C, 55, 69, 71, 72

J. Clegg 31, 52A,B

K.V. Frisch 30

M. Chinery 20A,B, 24B, 62G, 63A,C,D.

M. Tweedie 3, 11A,B,C, 14A,B,C, 18, 28, 41A, 57A, 61C, 73

P. Farb 9, 13B, 32, 33, 45, 46, 51, 58A,B, 59A,B,C

R. Goodden 1A, 2,5,6, 10A, 12A,B,C, 13A, 20D, 48, 49A,B, 68

R.H. Barth JR. 64, 65

R.J. Elzinga 20C, 29A,B, 60C, 61A,B, 62,B,C,D,E

T.M. Casey 59D,E,F,G,H

UNESCO 63B

V.B. Awasthi 62F, 66

W.M. Blaney 8, 15, 19E, 24A, 34A,B, 38, 39A,B, 50, 57B, 60B, 62A, 63F, 67, 70 A,B,C, 74, 75

W.R. Corti 1B, 10B, 21A,B,C

Other Related Book

Introduction to General and Applied Entomology (3rd Edition)

Scientific Publishers

V.B. Awasthi

This text book aims at to provide a comprehensive, well illustrated, up to date, general and applied account of Entomology which is written in a simple and lucid language keeping in view the syllabi of M. Sc. Zoology- Entomology and M.Sc. Ag.- Entomology of various Indian Universities. The book will also be useful to those who are preparing for competitive examinations conducted by various central and state government agencies for recruitment. The book is equally useful to those dealing with pests and pest control. It is hoped that the book will be useful not only to students but also to agriculturists, horticulturists, foresters, veterinary doctors and laymen dealing with pest control.

CONTENTS

ISBN: 978-81-7233-597-7, 2009, 499 pp., Hardback
978-81-7233-595-3, 2009, 499 pp., Paperback

Other Related Book

Scientific Publishers

Agricultural Insect Pests and their Control

V.B. Awasthi

This book has been mainly written keeping in view the requirements of the students of Agricultural Entomology of various Indian Universities. The book provides a brief account on the structure, function, development and metamorphosis, and a comprehensive account on almost all the methods of pest control. The nature of damage caused by the important insect pests of crops, their life histories and control have been described. It is hoped that the students of Agricultural Entomology, will find the book extremely useful. The book is also useful to the Plant Protection Departments and similar Institutions. The book will also be useful to those who are preparing for competitive examinations conducted by various central and state Government agencies for recruitment. Every attempt has been made to provide necessary information from the point of view of students of Agricultural Entomology, which is not available in other books.

CONTENTS

ISBN: 81-7233-491-5, 2008, 267 pages, Hardback
81-7233-493-1, 2008, 267 pages, Paperback

Other Related Book

A Manual of Practical Entomology

Revised & Enlarged 2nd Ed.

M.M. Trigunayat

A Manual of Practical Entomology (Field & Laboratory guide) is written to provide text material on different aspects of the practical syllabi of M.Sc. Entomology. All chapters are illustrative and well explained. Easy text will help students understanding the exercises. Dissections of the individual insect have been described so as to understand and follow the finest details of the anatomy of the insect concerned. Physiological and Behavioural Exercises are well presented usually in the style followed by the students. Individual Orders have been dealt with relevant necessary text and vital aspects of biology of the insect concerned. Chapters on Natural History and Rearing will inculcate interest among the students for their quest on these tiny creatures. Profusely illustrated with simple figures presentation and style gives the reader an insight to make it unique. Primarily intended for use by the post-graduate and graduate students of Entomology of the Universities in Rajasthan and elsewhere in India, it can also be used by agriculture departments, naturalists and workers in other related fields.

Contents

1. Insect Collection and Preservation; **2.** Classification of Insects; **3.** Survey of Representative Insect Orders; **4.** Identification of Insects with the help of Taxonomic Keys; **5.** Natural History of Common Insects; **6.** Applied Entomology; **7.** Exercises Based on Insect Physiology, Ecology Toxicology and Behaviour; **8.** Insect Anatomy; **9.** Histological Slides; **10.** Microtomy; **11.** Field Report; References and Suggested Readings; Appendix I; Appendix II; Appendix III; Subject Index.

ISBN: 978-81-7233-566-3, 2009, 351 pages, Hardback
978-81-7233-567-0, 2009, 351 pages, Paperback